Monitoring the Health of Populations by Tracking Disease Outbreaks

Saving Humanity from the Next Plague

ASA-CRC Series on
STATISTICAL REASONING IN SCIENCE AND SOCIETY

SERIES EDITORS
Nicholas Fisher, University of Sydney, Australia
Nicholas Horton, Amherst College, MA, USA
Regina Nuzzo, Gallaudet University, Washington, DC, USA
David J Spiegelhalter, University of Cambridge, UK

PUBLISHED TITLES

Errors, Blunders, and Lies: How to Tell the Difference
David S. Salsburg

Visualizing Baseball
Jim Albert

Data Visualization: Charts, Maps and Interactive Graphics
Robert Grant

Improving Your NCAA® Bracket with Statistics
Tom Adams

Statistics and Health Care Fraud: How to Save Billions
Tahir Ekin

Measuring Crime: Behind the Statistics
Sharon Lohr

Measuring Society
Chaitra H. Nagaraja

**Monitoring the Health of Populations by Tracking Disease Outbreaks:
Saving Humanity from the Next Plague**
Steven E. Rigdon and Ronald D. Fricker, Jr.

For more information about this series, please visit: https://www.crcpress.com/go/asacrc

Monitoring the Health of Populations by Tracking Disease Outbreaks

Saving Humanity from the Next Plague

Steven E. Rigdon
Ronald D. Fricker, Jr.

CRC Press
Taylor & Francis Group
Boca Raton London New York

CRC Press is an imprint of the
Taylor & Francis Group, an **informa** business

A CHAPMAN & HALL BOOK

CRC Press
Taylor & Francis Group
6000 Broken Sound Parkway NW, Suite 300
Boca Raton, FL 33487-2742

© 2020 by Taylor & Francis Group, LLC
CRC Press is an imprint of Taylor & Francis Group, an Informa business

No claim to original U.S. Government works

Printed on acid-free paper

International Standard Book Number-13: 978-0-367-24283-1 (Hardback)
978-1-138-74234-5 (Paperback)

Library of Congress Cataloging-in-Publication Data

Names: Rigdon, Steven E., 1955- author. | Fricker, Ronald D., Jr., 1960-, author.
Title: Monitoring the health of populations by tracking disease outbreaks : saving humanity from the next plague / by Steven E. Rigdon & Ronald D. Fricker, Jr.
Other titles: ASA-CRC series on statistical reasoning in science and society.
Description: Boca Raton : CRC Press, [2020] | Series: ASA-CRC series on statistical reasoning in science and society | Includes bibliographical references and index.
Identifiers: LCCN 2019056089 (print) | LCCN 2019056090 (ebook) | ISBN 9780367242831 (hardback) (alk. paper) | ISBN 9781138742345 (paperback)(alk. paper) | ISBN 9781315182384 (ebook)
Subjects: MESH: Epidemiologic Methods | Disease Outbreaks | Statistics as Topic | Public Health Practice | Biostatistics--methods
Classification: LCC RA652.2.M3 (print) | LCC RA652.2.M3 (ebook) | NLM WA 950 | DDC 614.4072/7--dc23
LC record available at https://lccn.loc.gov/2019056089
LC ebook record available at https://lccn.loc.gov/2019056090

Visit the Taylor & Francis Web site at
http://www.taylorandfrancis.com

and the CRC Press Web site at
http://www.crcpress.com

*Dedicated to all those who seek
to eradicate human disease.*

Contents

Preface

We live in an era in which modern medicine can cure many diseases. Seemingly gone are the days when deadly illnesses like the plague could sweep through communities with impunity killing millions. As a result, today the citizens of developed countries have never experienced a large-scale disease outbreak.

Most people would properly credit the successful control of deadly historical diseases to the truly amazing progress that has been made in the science and practice of medicine in the past century. These include advancements in the basic science of disease etiology, the development of antibiotics and other treatment drugs, and the creation of our modern medical and public health infrastructure.

Less visible to the general public, but no less important, is the role of the public health community – including epidemiologists and biostatisticians – in tracking and identifying disease outbreaks, which we define to be a sudden increase in the rate of a disease beyond the normally occurring rate. While we use the term *epidemic* to mean a widespread outbreak, we call a worldwide epidemic a *pandemic*. An important part of this process is identifying what causes a disease so that steps can be taken to control its spread. When it comes to disease outbreaks, the work of public health officials often critically depends on the use of statistical methods to help discern whether an outbreak may be occurring and, if there is sufficient evidence of an outbreak, then to locate and track it.

Given the success of public health interventions, someone casually reading this Preface might think this book is mostly historical and they no longer need to worry about massive disease outbreaks. For example, one might think the days of the 1918 Spanish Flu that killed somewhere between three and five percent of the world's population, or the "Black Death" bubonic plague of the 14th century that killed about a third of Europe's population, are long gone. But that would be wrong.

Consider that on November 16, 2019, Chinese authorities diagnosed a case of pneumonic plague in a 55-year-old man and quarantined 28 people who came in close contact with the man. The pneumonic plague is actually more contagious than the Black Death because it can be transmitted from person-to-person. Or consider the Ebola virus that continues to come close to evading containment by public health authorities. From 2014 to 2016, in just Sierra Leone, Liberia, and Guinea alone, more than 28 thousand people have been infected and more than 11 thousand have died. The latest outbreak in 2018 killed more than 2,000 people.

Also, as this book goes to press, the 2019 Novel Coronavirus (2019-nCoV) is challenging governments and health organizations to contain it. As of early February 2020, The World Health Organization has declared 2019-nCoV a "public health emergency of international concern" and travel bans to China are being implemented, borders are being closed, and quarantines are being implemented and enforced. It is too early to tell whether these measures will be fully effective or not.

Evidence currently suggests that 2019-nCoV is not as deadly as severe acute respiratory syndrome (SARS), which had a mortality rate of about 9 percent compared to current estimates of less than three percent for the coronavirus. However, more people have already died in China from 2019-nCoV than from SARS.

There are a number of key factors that drive how far and how fast a disease can spread and how deadly it will ultimately be. These include:

- How contagious the disease is, where the more contagious it is the quicker it can spread.

- How hardy it is, meaning how long the pathogen can live outside of the body and still infect those who come into contact with it.

- How long it takes to show symptoms and whether a person is contagious before showing symptoms, where diseases that are contagious before showing symptoms are easier to spread and it is harder to identify those who are contagious.

- How virulent the disease is, where the higher the morbidity and mortality of the disease the worse its impact.

The worst-case scenario is an extremely virulent, highly infectious disease that is contagious but asymptomatic for a significant period of time. Such a disease will be very hard to contain and will be devastating to the affected population. As of this writing, 2019-nCoV seems not be such a disease, where current information suggests it is only moderately infectious (though it is contagious when a person is asymptomatic for a period from 2 to 14 days) and the mortality rate is relatively low compared to a disease like Ebola that has an average mortality rate of around 50 percent. However, 2019-nCoV is a major public health threat challenging containment, and one day an even more contagious and more virulent disease *will* emerge and for that we need to be well and fully prepared.

So, this is not a history book. Just the opposite: it is a book focused on how statistics in the service of public health can be used so that we do not repeat our human history of large, virulent, deadly disease outbreaks.

This book is the story of the application of statistics for disease detection and tracking. It is the story of how medical and public health professionals use statistics to separate critical disease information from all the noise of our modern world so that they can most effectively intervene and mitigate the effects of the disease. While we mostly cover communicable diseases such as influenza and Zika, we also discuss surveillance of noncommunicable diseases such as cancer. These noncommunicable diseases often have an identifiable cause that, if found, could lead to appropriate action and a reduction in the health risk.

The book is divided into two parts. The first part talks about the concepts and methods of disease surveillance. This part begins with the issue of separating the signal (a true change in the rate of a disease) from the random noise that is ubiquitous to disease monitoring, as well as other areas where statistics is applied, such as the control of manufacturing or service quality. We discuss the types of surveillance that are often done to monitor disease, which includes not just the monitoring of human disease, but also the surveillance of animals and agriculture. Surveillance on humans can be divided into traditional surveillance, where the numbers of confirmed cases is tracked, and syndromic surveillance, where symptoms of diseases (not confirmed cases of the disease) are tracked. Recently, epidemiologists have used indirect approaches to track disease. These include internet searches, pharmaceutical sales, futures markets, and others.

The second part of the book is about disease investigations. We begin with a description of the steps an epidemiologist might take in an investigation. The epidemiologist acts much like a detective in collecting and evaluating evidence in order to reach a conclusion about a case. The next seven chapters describe seven actual disease investigations which include: the Nipah virus, smallpox, syphilis, anthrax, cancer, yellow fever, and microcephaly (and its relationship to the Zika virus). The story of the 2019-nCoV investigation is unfolding in real time as this book goes to press and may one day be a chapter in some future book, a chapter that we sincerely hope tells a story of successful containment and eradication of 2019-nCoV.

We hope the reader sees how statistics plays an important role in disease surveillance. Separating the signal from the noise is just one part of the story. Statisticians also help design surveys and experiments that will yield the information that is needed. Statisticians then analyze the resulting data to help the investigators zero in on a cause for a disease. Through the process of identifying a disease outbreak, finding its cause, and developing a plan to prevent its reoccurrence, statisticians and epidemiologists can improve public health across the world.

Authors

Dr. Steven E. Rigdon is a professor in the Department of Epidemiology and Biostatistics in the College for Public Health & Social Justice at Saint Louis University. He holds a PhD and an MA in Statistics from the University of Missouri-Columbia, as well as an MA and BA in Mathematics, from the University of Missouri-St. Louis. He is also Distinguished Research Professor Emeritus in the Department of Mathematics and Statistics at Southern Illinois University Edwardsville.

Author of several books, including *Statistical Methods for the Reliability of Repairable Systems* published by John Wiley & Sons, and *Calculus*, 8th ed. published by Pearson, Dr. Rigdon has published more than 80 peer-reviewed journal articles and book chapters. He is a Fellow of the American Statistical Association (ASA) and a member of the International Society for Disease Surveillance. He is also editor of *Journal of Quantitative Analysis in Sports*. In his spare time, Dr. Rigdon plays the French horn and trumpet and he is an ice hockey official.

Dr. Ronald D. Fricker, Jr. is the Associate Dean for Faculty Affairs and Administration in the Virginia Tech College of Science. He is also a professor in the Virginia Tech Department of Statistics and is a past head of the department. He holds a PhD and an MS in Statistics from Yale University, an MS in Operations Research from The George Washington University, and a bachelor's degree from the United States Naval Academy. Author of *Introduction to Statistical Methods for Biosurveillance* published by Cambridge University Press and nearly 100 papers, monographs, reports, and articles, Dr. Fricker is Fellow of the ASA and an Elected Member of the International Statistical Institute. He is a former chair of the Section on Statistics in Defense and National Security and a former chair of the Committee on Statisticians in Defense and National Security, both of the ASA.

CHAPTER 1

The Next Plague

IMAGINE a pair of ecotourists from New York who travel to Southeast Asia and, while visiting a live meat market, come into contact with chickens carrying a new strain of avian (bird) flu. One of them acquires the virus at the market, and the other acquires it in the next day through person-to-person contact. Under such a scenario, the infected tourists could fly from Bangkok to Hong Kong with 300 or more airline passengers and perhaps come into direct contact with 30 of them plus several of the flight crew. Maybe half of the 30 will develop the avian flu. When the plane reaches Hong Kong, most people will go their separate ways: stay in Hong Kong or continue travel to cities like Tokyo, Sidney, San Francisco, Warsaw, or London. Although it is unlikely that these newly infected passengers would further infect other passengers, there would be people all over the world who have contracted the disease and eventually become infectious. Then when the ecotourists fly from Hong Kong to London, and from London to New York, they would come into contact with two additional planeloads of passengers. After all this travel, it would be almost certain that within two days every continent except Antarctica would contain people infected with the disease.

When the original pair of infected tourists reach home, they would feel sick for a few days, but would likely seek medical treatment first from their physician. Because any new strain of avian flu is likely to produce symptoms that mimic the seasonal flu, it is possible that a physician will diagnose the couple as having the seasonal flu. The physician has probably seen at least a dozen cases of apparent seasonal flu that week and diagnosed and treated them for that. It could well be a few more days, when the symptoms, such as respiratory distress, become more pronounced that the couple presents at a hospital's emergency department. When more detailed inquiries are made, especially regarding the international travel, a more accurate diagnosis could be rendered. New diseases such as this often take weeks or longer before enough people contract the disease to give it a name, a description of symptoms, and suggested treatment.

This scenario of flu misdiagnosis is likely to play out in cities across the world with those who were unfortunate enough to come into physical contact with our ecotourists. Since most influenza viruses have an incubation period of two to five days, by the time anyone receives a diagnosis the number of sick people could be in the hundreds and the number who are infected but not yet sick could be in the thousands. If the bird flu is not caught early, it wouldn't take very long for what started as a small outbreak to spiral into a worldwide epidemic, called a *pandemic*.

A Closer Look

Throughout history, the human race has periodically been ravaged by disease. The 1918 "Spanish flu" pandemic, the first of two pandemics involving the H1N1 virus, infected 500 million people worldwide and killed between 50 and 100 million, or about three to five percent of the entire world's population. The bubonic or Black Death plague pandemic of the 14th century is estimated to have killed as many as 25 to 30 million people in Europe; in some locations such as Florence it killed nearly six out of every ten people.

While we often think about plagues as ancient history, recent disease outbreaks include:

- H1N1 "swine flu" pandemic, which was first detected in the United States in April 2009, and from April 12, 2009 to April 10, 2010 resulted in approximately 60 million flu cases, almost 275,000 hospitalizations, and more than 12,000 deaths in the United States alone.

- West Nile virus, which is transmitted by mosquitoes, arrived in the US in 1999 and through 2015 infected 43,937 and killed 1,911 people in the US.

- Severe acute respiratory syndrome (SARS), which is transmitted person-to-person via sneezing, touch, or other close contact, infected 8,098 people worldwide in 2003 of which 774 died.

- Zika virus, transmitted primarily by the *Aedes aegypti* mosquito (see Figure 1.3), can cause congenital malformations in infants when mothers are infected during pregnancy. The first recorded Zika outbreak was on the Island of Yap (in the Pacific Ocean, east of the Philippines) in 2007, followed by a large outbreak in French Polynesia in 2013, and then one in Brazil in 2015. As of 2018, 86 countries and territories have reported evidence of mosquito transmission of the Zika virus. In the US there have been 231 cases "acquired through presumed local mosquito-borne transmission," mostly in 2016 (224) and in that year, most were in Florida (218).[1]

- The Ebola virus, which most recently, came close to evading containment by public health authorities. In just Sierra Leone, Liberia, and Guinea, more than 28,000 people were infected of which more than 11,000 died from 2014 to 2016.

- In late 2019 a Novel Coronavirus, called 2019-nCoV, emerged in Wuhan, China. As of early February 2020, the virus had sickened over 30,000 and killed almost 700. In response, quarantines and restrictions on travel from China are being implemented. Time will tell how effective these measures are.

1.1 UNDERSTANDING PANDEMICS

Thankfully, the story of the ecotourists is just that: a story based on a hypothetical scenario. But it is not an unrealistic story or some far-fetched Hollywood movie plot that could never occur in real life. It is a story that very much could occur and, in fact, may even have come close to actually occurring.

How can we know that? Well, one type of bird flu, designated H5N1,[2] has infected more than 850 people, mainly in Southeast Asia and Africa, and it has resulted in more than 450 deaths – a fatality rate greater than 50 percent; this death rate is much higher than most other forms of influenza.[3]

Another bird flu strain, H7N9, has infected more than 600 people and, of them, more than 200 have died – a fatality rate greater than 33 percent. What separates our hypothetical scenario from reality is that – thus far – neither the H5N1 nor H7N9 (nor any other) bird flu strains have mutated in such a way that they are easily transmissible between people.[4] Should that ever happen, then this hypothetical scenario would no longer be hypothetical; it will have crossed over from being a scary story into becoming a real-world nightmare.

So, you might be thinking, "Okay, but how is it possible that thousands of people could end up being exposed to the bird flu in only one week after the first person becomes contagious?" To get some insight, let's look at a very simple example, where each person who is exposed to the bird flu becomes contagious and infects exactly five other people the next day (and then ceases to be contagious). Each day, the number of new infected people is five times that of the previous day. Over the course of seven days, we have

$$5 \times 5 \times 5 \times 5 \times 5 \times 5 \times 5 = 5^7 = 78,125 \text{ infected people.}$$

In this situation, the number of infected people increases fivefold *every day*. As illustrated in Figure 1.1, like a tree with ever more branches, this is an exponentially increasing phenomenon. "Patient 0" or the "index patient" infects five other people, each of whom infect five more apiece, etc. If the contagion process continues for just three more days the 78,000 infected becomes nearly 10 million infected. Just two more

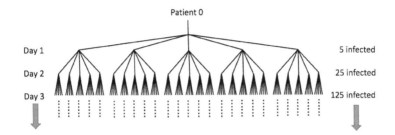

Figure 1.1 In the absence of an intervention, this tree illustrates how the number of infected people grows exponentially. By Day 3 the number of infected is 125; in 7 days it grows to over 78,000 and to 10,000,000 in 10 days.

days, and the number reaches nearly a quarter of a billion people!

Statistically Speaking

Starting with one unit of "x," the formula for calculating the total quantity of x at time n (denoted as x_n) with growth rate r is

$$x_n = 1 + r + r^2 + \cdots + r^{n-1} = \sum_{t=1}^{n} r^{t-1} = r^n - 1,$$

where time is measured in discrete intervals ($t = 1, 2, 3, \ldots, n$). The surprise of exponential growth has been illustrated via stories over the ages. One dating back to the 12th century has a king asking a courtier what he would like in return for the gift of a beautiful chessboard. The courtier tells the king he would like a quantity of rice calculated as follows: put one grain of rice on the first square of the chessboard, two grains on the second, four grains on the third, eight on the fourth, etc. Each day, the number of grains is double that from the previous day. The king agrees only to find out there is not enough rice in the whole world to fulfill his bargain:

$$x_{64} = 2^{64} - 1$$
$$= 18,446,744,073,709,551,615 \text{ grains of rice.}$$

At an average of 50 grains of rice per gram, this is more than 370 *billion* metric tons of rice. To put this in context, in 2017, the United States Department of Agriculture estimated that the

2017 worldwide production of rice was 483 *million* metric tons. The required amount of grain is more than 700 times the current world production!

As this simple example makes clear, some important characteristics of a pandemic include: (1) how quickly someone who is exposed to the disease becomes contagious, the disease incubation time, and (2) how long it takes to discover the existence of the pandemic. If in this example it takes a week to become contagious rather than one day, then at the end of the week there are only five infected people rather than 78,000. Similarly, if the disease is easily discoverable after two days then even if people become contagious after one day, the pandemic can be identified after only $5 \times 5 = 25$ people are infected.

Other critical pandemic characteristics are the rate at which contagious people expose others to the disease and how infectious the disease is. The bird flu story is, in some ways, a worst-case scenario because the index patient exposed a large number of people to a highly infectious bird flu as she traveled back to New York City. Suppose for example that she came into contact with 217 people, and about three-fourths (say 164) of them contracted the flu and subsequently became contagious. Assuming a 24-hour incubation cycle and that each contagious person after the index patient infects exactly five other people, the number of contagious and sick persons is shown in Table 1.1.

What we see in the table is that highly contagious diseases can spread very rapidly. With a 50 percent fatality rate after eight days of illness, the number of deaths starting on Day 10 will climb dramatically, very rapidly reaching into the thousands and then millions. By Day 13, the number of fatalities will be 21,939. The numbers in Table 1.1 grow so rapidly that the number of exposed persons by Day 13 is a significant fraction of the world's population. A more accurate model would account for the finite size of the world's population and the

Table 1.1 Daily numbers of people who are contagious, exposed, and die in the hypothetical bird flu example. Except for Day 0, the number contagious on day i is 75 percent of those exposed on day $i - 1$. And, the number exposed on day i is five times the number contagious on day i.

Day	Contagious	Exposed	Deaths
Day 0: Sunday	1	217	
Day 1: Monday	163	815	
Day 2: Tuesday	611	3,055	
Day 3: Wednesday	2,291	11,455	
Day 4: Thursday	8,591	42,955	
Day 5: Friday	32,216	161,080	
Day 6: Saturday	120,810	604,050	
Day 7: Sunday	453,038	2,265,190	
Day 8: Monday	1,698,893	8,494,460	1
Day 9: Tuesday	6,370,845	31,854,225	82
Day 10: Wednesday	23,890,669	119,453,345	306
Day 11: Thursday	89,590,009	447,950,045	1,146
Day 12: Friday	335,962,534	1,679,812,670	4,296
Day 13: Saturday	1,259,859,502	6,299,297,510	16,108
TOTAL			**21,939**

fact that eventually the virus will run out of healthy people to infect.

Of course, the "starting conditions" matter a lot, where in this scenario the initial exposure of 217 people drives the incredibly large number of people exposed. On the other hand, the scenario was actually quite conservative in assuming that each contagious person after the index patient exposed only five other people. Think of how many people you interact with on a daily basis.

So, once again, you might be thinking, "Fine, so the math in this simple example results in a huge outbreak, but could that happen in the real world? Aren't these numbers just massively overblown?" Sadly, there is clear historical evidence that pandemics this large have happened. The Spanish flu of 1918 to 1920 infected half a billion people at a time when the world's population was only 1.5 billion. Estimates of the number who died range from 20 to 50 million. This was at a time when there was no commercial aviation and very little automobile travel; railroad access was limited for much of the world's population. Imagine if an influenza virus similar in effect to what caused the Spanish flu emerged today when people can travel virtually anywhere in the world in 24 hours.

A Closer Look

Quarantine, *isolate*, *inoculate*, and *vaccinate* are strategies used to contain an outbreak. Although they achieve the same objectives, they are different.

Isolation is the practice of separating sick persons from healthy persons in an attempt to restrict the transmission of the disease from the sick to the healthy.

Quarantine is the practice of restricting the movement of healthy persons who have been or might have been exposed to a disease, thereby preventing the spread of a disease.

Inoculation produces immunity in a person by injecting a form of a disease to prevent a more serious form.

Vaccination refers to the practice of injecting germs to prevent a disease. For example, an injection of cowpox is sufficient to prevent smallpox (a much more serious disease). This would be referred to as a cowpox inoculation or a smallpox vaccine.

As we will explore in the next section, how communicable diseases spread among a population is actually much more complicated than our simple examples here. However, the key point is that it is critically important to detect highly infectious, highly virulent diseases as early as possible, and then to track down and inoculate (or quarantine) every single exposed person.

For example, in October of 2017 Madagascar had an outbreak of the bubonic plague, called the Black Death in the

Middle Ages. The bubonic plague is actually not all that rare in Madagascar, where there were about 300 cases in the month of October for each of 2015 and 2016. However, in the three months leading up to October of 2017 the number of bubonic plague cases jumped to more than 2,000.

These days, bubonic plague is easily cured with antibiotics, but left untreated can develop into pneumonic plague which easily spreads from person to person and is fatal. By October of 2017, 171 people of the 2,000 infected had died in Madagascar. To control the outbreak, public health workers tracked down more than 7,000 people who had interacted with those who were either confirmed or suspected of contracting the plague, eventually treating about 9,300 people with antibiotics. As Figure 1.2 shows, an outbreak that was growing rapidly in early October was quickly contained and largely brought under control by the end of the month.

1.2   MODELING PANDEMICS

What the Madagascar outbreak shows is that today, once an outbreak is identified, public health resources are marshaled and applied to mitigate and then eliminate the outbreak. What this means is that modeling an outbreak is typically not as simple as the bird flu scenario, where the model must account for the interplay between the human population, perhaps an animal host (such as fleas that carry bubonic plague), the disease, and public health intervention.

One common way to model disease spread is to use a *compartment model* that abstracts the population into a set of compartments into which persons are classified according to their health status with respect to the disease. The SIR model is such a model, with three compartments, where the "S" stands for *susceptible*, the "I" for *infected*, and "R" for either *recovered* or *removed*.

In the simplest case, known as the Kermack-McKendrick model and illustrated in Figure 1.4, the spread of the disease

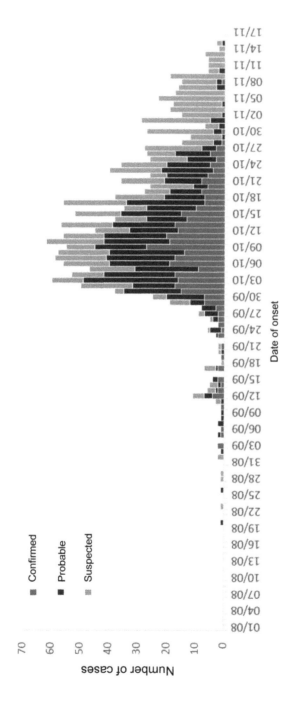

Figure 1.2 The number of confirmed, probable, and suspected cases of pneumonic plague in Madagascar from August 1 ("01/08") through November 17, 2017 ("17/11").[5]

Figure 1.3 The *Aedes aegypti* mosquito. © Shutterstock Images.

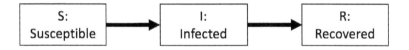

Figure 1.4 The basic SIR model.

is governed by two parameters: β, the infection rate, and γ, the recovery rate. The model assumes:

- the size of the population is fixed,

- the disease happens instantaneously (i.e., there is no incubation period),

- the population is homogeneous, and

- those who have recovered are immune to reinfection.

The way the disease spreads is then defined by three equations, where:

- the number of susceptible individuals decreases as a function of the infection rate times the number of susceptible people (S) and the number of infected people

(I). The idea is that the interaction between the susceptible people and infected people drives the infection rate
according to how infectious the disease is;

- then, the number of infected people increases according
to the number of susceptible people who get infected
from the first equation minus the number who recover;
and,

- finally, the number of people who recover is the product
of the number of infected individuals times the recovery
rate.

Of course, the assumptions and the equations are simplifica-
tions of the real world, but even such a simple model can provide useful insights into how to successfully fight an outbreak.
With such a model, researchers can explore public health policy options to mitigate the spread of a disease. For example,
an influenza outbreak could be attacked by doing some combination of the following:

- Recommend the widespread use of antiviral drugs. This
would have the effect of decreasing the duration of the
flu, thereby increasing the recovery rate γ.

- Recommend getting a flu shot. Getting the flu shot
would effectively move people from the susceptible group
to the "recovered" group.

- Implement an education campaign. For example, "Stay
home if you're sick!" or "Wash your hands frequently."
This has the effect of decreasing the rate β of spread of
the disease.

Figure 1.5 shows how the model would predict the effect
of each scenario, along with the "do nothing" strategy in the
upper left. The SIR model assumes $\beta = 3.6$ and $\gamma = 3.2$, which
are reasonable for influenza. With no intervention, there would
be about 228,000 cases. The most effective strategy is the flu
vaccine, where the number would be reduced to 103,000 cases.

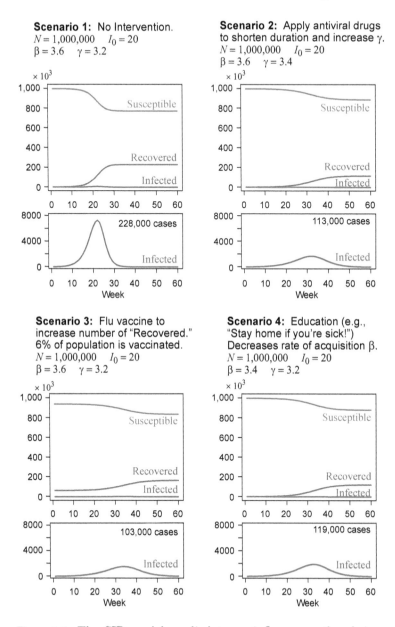

Figure 1.5 The SIR model applied to an influenza outbreak in a city of 1,000,000 people, where initially $I_0 = 20$ people have the disease. The blue curves indicate the number of those susceptible, the red curves show the number infected, and the green curves indicate the number recovered. The number of cases of influenza is shown for each of the four scenarios.

Even if the SIR is not a perfectly accurate model (no models are) they often allow epidemiologists to play "what if" games to see the effect of various scenarios. To explore this further, we could ask "what if β could be reduced to 3.5, and γ could be increased to 3.3, and 4% of the population got a flu shot?"

Returning to the bird flu example, the SIR model is not particularly realistic for this scenario. One reason is that there is an incubation period during which the individual has been infected but is not yet contagious and it is possible for someone who has recovered to then be susceptible to reinfection. Figure 1.6 shows a more appropriate SEIR model in which the "E" compartment allows for those individuals who are exposed but not yet infected (i.e., contagious). It also allows for those who are exposed but not infected and those who are infected and recovered to return to the susceptible compartment. Finally, note that the "R" compartment now includes those who have recovered as well as those who have been removed (which is a euphemism for those who have died from the infection).

Compared to the SIR model, this SEIR model requires more parameters and equations describing how the disease progresses. These models work best when a disease spreads from direct contact; when this holds, the models' assumptions seem reasonable. Many other diseases do not spread through direct person-to-person contact so adjustment of one of these models may be necessary. For example, some diseases are carried by an intermediary, such as a mosquito. While there is no direct contact between persons, it is still true that the more people with the disease, the more likely it is that a

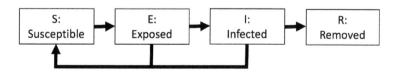

Figure 1.6 An SEIR model for the bird flu scenario.

mosquito will bite an infected person, and subsequently an uninfected person, thereby spreading the disease. As we'll see in the next section, some diseases are spread by polluted water. We needn't get into all the details here except to note that these types of models can be made as complex as the modeler desires, though the more complicated the model the more difficult it can be to determine the model parameters. Indeed, sometimes more complicated does not necessarily mean better when it comes to modeling.

1.3   JOHN SNOW AND CHOLERA

The use of statistical analyses to help determine the origin of disease goes back to the Soho (an area on the west side of London) cholera pandemic of 1839 to 1856 in England where Dr. John Snow, a London physician, mapped cases of cholera to support his theory that the disease was transmitted by contaminated water. Today we know that cholera is caused by the bacterium *Vibrio cholerae* that is most often transmitted through water (and sometimes food). However, in the mid-1800s, how cholera spread was not understood. Some thought it was spread contagiously from person to person and others thought it was attributable to some sort of environmental pollution. Because the disease often disproportionately affected the poor, the biological aspects of the disease were also confounded with questions of morality where, particularly in the United States, cholera was often attributed to God's wrath for an individual's or group's moral failings.

Figure 1.7 is a portion of Snow's now famous plot simultaneously showing the number of cholera deaths by city address and the locations of each of the city's water pumps. Pump locations are depicted by dots in middle of the streets and the bars are proportional to the number of deaths at each address. In the center of this map excerpt is the Broad Street pump and in the larger map there is a clear visual association between areas of higher death rates from cholera and specific water pumps, particularly the Broad Street pump.

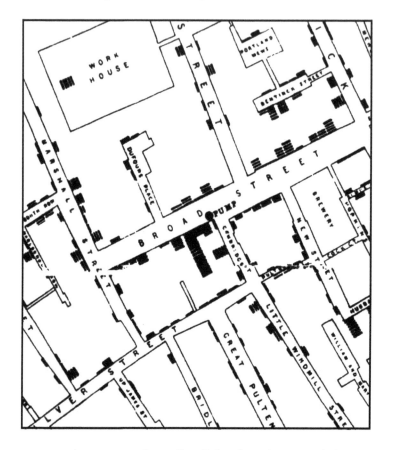

Figure 1.7 An excerpt from Dr. John Snow's map of the 1854 London cholera epidemic.[6] The Broad Street pump, which turned out to be the cholera source, is at the center. Each bar corresponds to one death at a particular street address.

When Snow made his map in 1854, the world was experiencing its third cholera pandemic. The first began in India in 1817 and by 1824 spread through much of Asia, Indonesia, and the Middle East. The second began in 1827, also in India, but this time spread through Southern Asia and the Middle East to Europe and eastern North America. The epidemic continued until 1835.

In a dramatized version of history, it is often told that John Snow created the map in Figure 1.7, then took the handle off the Broad Street pump to demonstrate that polluted water was the way cholera was spread, and the health problem was solved. Reality is much more complicated. The epidemic was waning when Snow removed the pump handle, so it is difficult to ascertain its effect; the epidemic may have come to an end without the intervention.

Snow concluded that drinking polluted water, not direct person-to-person contact, caused cholera. More infected persons led to more human waste entering the water system, which led to more people drinking the water that was loaded with *Vibrio cholerae*. Thus, cholera was spread from person to person, but in an indirect way.

The third pandemic started in 1839 and spread even farther to the west in North America and to South America. Although Dr. Snow did convince the local governing board to remove the pump handle, there was no consensus, or even real understanding in a medical sense, of the causes of cholera. Indeed, it wasn't until the fourth epidemic (1863–1875) that Snow's theory that the disease was spread via infected drinking water started to truly gain traction. Even though the importance of sanitation was becoming clear, there was still great uncertainty about the actual cause of cholera.

It was not until the fifth cholera pandemic from 1881 to 1896 that the German physician and microbiologist Robert Koch* (1843–1910) identified the *Vibrio cholerae* bacteria as the potential cause, but he was unable to produce cholera symptoms in animals. There then followed a sixth cholera pandemic (1899–1923), by which time Western physicians and public health professionals began to accept that cholera was caused by *Vibrio cholerae* and had developed public health policies to effectively combat the disease.[7]

---

*Koch was awarded the Nobel prize in physiology and medicine in 1905, not for his work on cholera, but for discovering that tuberculosis was caused by a bacterium called *Mycobacterium tuberculosis*.

Much of the difficulty in determining the cause of cholera had to do with the lack of scientific knowledge and tools of that era. However, even in the modern era the cause of diseases like AIDS or microcephaly (infants born with abnormally small heads and brains) can be hard and take time, effort, and creative experimentation to determine. So today, just as in the mid-1800's, epidemiologists and public health professionals are called upon to identify and then mitigate disease outbreaks, much as Dr. Snow first did almost two centuries ago. When they do so, they now automatically turn to data and the use of statistical methods to find and track outbreaks, perhaps well before the science behind the cause of the disease and the medical solution for it have been found.

1.4   ABOUT THE REST OF THIS BOOK

It is impossible to tell when the next pandemic will occur or where it might originate. The widespread use of international air travel and its speed mean that the disease could spread around the world in a short period of time, possibly before it was even noticed or diagnosed.

Since most viral diseases are *zoonoses*, which means they can infect animals in addition to humans, and they can spill over from one to the other, the most likely scenario is that humans are initially infected from an animal, such as a bird. (In some cases, the animal may be unaffected by the virus but will still carry and potentially spread the virus to other animals or humans.) A slight mutation could alter how a virus affects humans and how humans react – or fail to react – to treatment.

This book is about the application of statistical methods for disease detection and how epidemiologists use statistics to discover the source of an increase in some disease and then mitigate its effect on people. This is the story of how, particularly in the early stages of a communicable disease outbreak, medical and public health professionals use statistics to separate critical outbreak information from all the noise in disease data so that they can most effectively discern when and where a disease outbreak may be occurring and can work to mitigate the effects of the outbreak. These are critical tools used by epidemiologists and biostatisticians, where their work often critically depends on the use of statistical methods to help to track, identify, and control the spread of disease.

Notes

[1] See the CDC website https://www.cdc.gov/zika/reporting/2016-case-counts.html

[2] Influenza viruses have spikes called the haemagglutinin and the nuraminidase, onto which an influenza virus will attach itself. There are various kinds of haemagglutinin and neuraminidase and the pair determine the strain of influenza. For example, H7N9 represents the seventh type of haemagglutinin and the ninth type of nuraminidase. Altogether, there are seventeen types haemagglutinin (H) and nine types of neuraminidase (N).

[3] https://www.who.int/influenza/human_animal_interface/2019_06_24_tableH5N1.pdf?ua=1

[4] The scenario has been simplified to assume a 24-hour incubation time. Current data suggest an incubation period from two to five days for H5N1 and an average of five days for H7N9. Source: World Health Organization Avian and Other Zoonotic Influenza fact sheet updated November 2016. Accessed at http://www.who.int/mediacentre/factsheets/avian_influenza/en/ on November 25, 2017.

[5] Source: World Health Organization, *Plague Outbreak Madagascar, External Situation Report 12*, dated November 20, 2017 (http://www.afro.who.int/health-topics/plague/plague-outbreak-situation-reports).

[6] Source: http://en.wikipedia.org/wiki/1854_Broad_Street

[7] For additional details about the various cholera outbreaks, see Hays, J.N. (2005). *Epidemics and Pandemics: Their Impacts on Human History*, ABC-CLIO, Inc.

I

Disease Surveillance

In an effort to detect diseases before they become widespread, public health professionals monitor the incidence or prevalence of many diseases and syndromes. As the following chapters will describe, these public health surveillance systems range from requirements for medical doctors to immediately report a single diagnosed case of certain diseases to computer-based systems that analyze non-specific symptoms looking for unusual patterns across entire populations.

Some diseases, such as smallpox or anthrax, are so rare that even a single case is evidence of an outbreak, and a cause for concern. Many other diseases, such as influenza, pertussis (whooping cough), hepatitis C, etc., occur throughout the year. The normally occurring rate when there is no outbreak is called the *background noise*, or just simply the *noise*. When a disease begins to spread beyond this level we have a *disease outbreak*. The term *signal* is used to describe the extent of the disease rate increase over the noise.

A key focus of this section is the idea of separating a disease signal from noise in data. This could be at an individual level in which a doctor seeks to reach a diagnosis based on laboratory tests that are imperfect to determining whether a communicable disease outbreak is occurring amid the random increases and decreases of disease incidence in a population. In many ways, it comes down to the question of whether some observed uptick in disease incidence today is the start of an increasing incidence trend or just a "blip" that will then drop back down tomorrow or the next day.

An important aspect of communicable disease surveillance is early detection, where catching an outbreak as early as possible is critical to helping contain what might otherwise become a pandemic. To do this, particularly when looking at non-specific syndrome data, the challenge is developing and properly employing sophisticated statistical methods that are good at identifying increasing disease incidence trends while minimizing the chance that some temporary blip is incorrectly identified as a trend.

CHAPTER 2

Separating Signal from Noise

T HE goal in public health surveillance is to identify disease outbreaks in some population, whether human, animal, or plant, as early as possible. There are a number of reasons why this is a challenging problem, starting with the fact that we never know in advance which disease will afflict which population at what point in time. Even in the case of the seasonal flu, where we know people will be afflicted by it every winter, we don't know which strain or strains of the flu will predominate, which parts of the population will be most affected, and even when the flu season will start.

From a statistical point of view, the problem is compounded because the data are noisy: it varies from day to day and season to season for many reasons. These include the variation in the number of people afflicted as well as differences in the number of afflicted people we actually observe (since, for example, not everyone who gets the flu goes to see a doctor) to various types of measurement errors (such as misdiagnoses) and other sorts of data errors. This requires the employment of various types of sophisticated statistical methods so that medical and public health professionals can "see through" the noisy data to understand what is actually happening so they can take appropriate action.

2.1 WHAT IS "SIGNAL" AND WHAT IS "NOISE"?

Statisticians often talk about separating the signal from the noise in data. The signal is the underlying truth that is of interest in a particular problem while the noise is the inherent randomness that obscures the signal to a greater or lesser extent. A simple way to think about it is

$$\text{observed data} = \text{signal} + \text{noise}.$$

If there is no noise, there would be no randomness in the data, and the data we observe is exactly the signal we're interested in. (The problem becomes simple in this case.) However, the larger the noise is (relative to the signal), the harder the problem of trying to discern the signal becomes.

Statistically Speaking

Statisticians have a precise description of what a *stable* process is: a process is stable if the outcome values are independent (that is, today's value does not influence tomorrow's or any other day's output) and the probability distributions of the output are the same for each time period. This may be a bit restrictive (especially the independent part) but it gives some intuition about what noise means in our context.

The "seasonal flu" actually occurs year-round, but it is most common in the fall and winter. However, the start of the flu season can vary significantly, where it can be as early as October or as late as May. Outside of the months from December to March, the number of influenza cases per day (or week) roughly satisfies the description of stable given above.

The CDC compiles data on "peak month of flu activity," defined as "the month with the highest percentage of respiratory specimens testing positive for influenza virus infection during that influenza season." From 1983 to 2016, the peak month that occurred most often was February (14 seasons), followed by December (7 seasons), then March (6 seasons), and January (5 seasons).[1]

The CDC's FluView provides weekly influenza surveillance information in the United States via real-time visualizations of influenza information collected by CDC's monitoring systems. To see what is happening across the United States, in your region, or in your state, see www.cdc.gov/flu/weekly/fluviewinter active.htm.

To illustrate the problem, consider a simple example in which the flu incidence is normally constant at one percent of the population and then, when the annual flu season hits, it begins to increase linearly for some time. Let's consider two scenarios: (1) the flu incidence increases only a small amount (a light flu season), and (2) it increases a lot (a heavy flu season). Figure 2.1 shows what the signal looks like in these two scenarios, where the vertical axis is the disease incidence rate in term of the percent of the population with the flu and the horizontal axis is time measured in weeks (with "week 1" being the first week in January, etc.). Without noise, it is easy to detect when the outbreak occurs; in either the top or bottom graph in Figure 2.1 we can discern that the outbreak begins in week 8. Note that the incidence rate is never 0, so some people are getting the flu all year round.

Now, what does the signal look like when we add in some noise? Again, let's consider two scenarios: (1) a small amount of variability in the fraction of people who come in for treatment for the flu (low noise) versus (2) a lot of variability of the number who come in for treatment (high noise). Figure 2.2 shows what the resulting data would look like for the four possible combinations of signal and noise.

As we see in Figure 2.2, our ability to see the signal through the noise is a function of the relative magnitude of the signal compared to the noise. In Figure 2.2, the top two plots correspond to the light flu season (i.e., the left plot from Figure 2.1) but with either low noise (left) or high noise (right). Similarly, the bottom two plots correspond to the heavy flu season (i.e., the right plot from Figure 2.1) but with either low noise (left) or high noise (right).

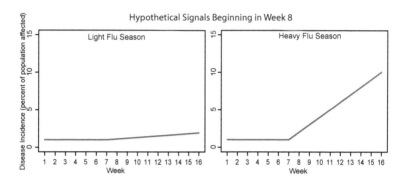

Figure 2.1 Hypothetical flu scenario "signals," where in the light flu season scenario (left plot) flu incidence increases by 0.1 percent each week starting in week 8, while in the heavy flu season scenario (right plot) flu incidence increases by one percent each week, also starting in week 8.

For example, while the signal is still discernible in the low noise scenarios on the left side, it is difficult to see for the light flu season. Remember, in the real world we don't observe the signal alone; we can only observe signal plus noise. Indeed, if we didn't already know from Figure 2.1 that an outbreak was occurring in the light flu season, there is a good chance we would not consider the incidence rates higher in weeks 8 through 16 compared to weeks 1 through 7. Furthermore, in the high noise scenarios on the right, the signal is impossible to see in the light flu season and even difficult in the heavy flu season. In fact, in the light flu season with high noise, one might even incorrectly conclude that the flu incidence rate is decreasing over the 16 weeks, which *is* true for this particular set of data just because of the random fluctuations due to the noise, but it is *not* true of the underlying signal itself.

Unfortunately, the ebb and flow of disease incidence is not as simple as in this hypothetical flu example, where it was a constant one percent until the flu season. Instead, disease incidence tends to increase and decrease over time as a result of many factors, some systematic (such as people tend to see

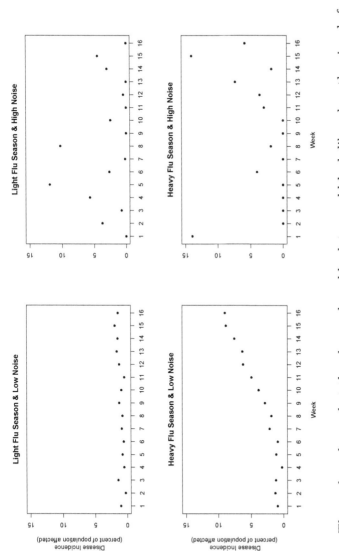

Figure 2.2 These plots show what the observed weekly data would look like when the signals from Figure 2.1 now have added noise (low and high). The signal is still discernible in the low noise scenarios on the left side, though it is difficult to see for the light flu season. In contrast, in the high noise scenarios on the right, the signal is impossible to see in the light flu season and even difficult in the heavy flu season.

their doctors more on the weekdays than weekends) and some just the result of a variety of naturally and normally varying factors. Key is separating out these "normal" trends, which we may not be interested in detecting (i.e., they are not the signal we are looking for) from those that we are (such as unusual increases signaling a potential outbreak).

In a more statistical treatment of the subject, we would delve into these issues in much more detail.[2] For our purposes, suffice it to say that it is a very challenging problem both substantively and statistically. Substantively it is difficult to precisely define what "normal" is and then what constitutes an outbreak. Statistically, it is very difficult to specify appropriate quantitative methods for both modeling "normal" and detecting outbreaks in the absence of precise definitions, and there are many technical challenges beyond the scope of this treatment.

2.2 QUANTIFYING THE NOISE

For our purposes, let's leave most of these challenges to the statisticians and epidemiologists. Here we'll just focus on how one might quantify the noise in data. The measure most commonly used is the *standard deviation*, which is typically denoted by the letter s. As shown in the "Statistically Speaking" section on page 29, the standard deviation formula looks complicated, but really it is just a way to quantify the variability in data in terms of a single number that expresses how spread out the data are around the average of the data.

To develop some intuition about how it works, let's return to counting people who go to 30 hospitals with flu-like symptoms. Figure 2.3 shows some plots of data that all have the same average (100) but different standard deviations ($s = 1.11, 4.99, 15.97$, and 33.78). As we see, the larger the standard deviation the more spread out the data are around the average (shown as the dotted line at 100). For example, in the upper left plot with a standard deviation of just over one, the data range from a low of 98 to a high of 102. In contrast, in the lower right plot with a standard deviation of almost 34, the data range from a low of 33 to a high of 186.

Statistically Speaking

Imagine our data consists of n observations, where perhaps each observation is the number of people who went to one of n hospitals with flu-like symptoms. We could denote these daily counts generically as $x_1, x_2, x_3, \ldots, x_n$. Then the average, which we denote as $\bar{x}$ (spoken as "x-bar"), is calculated as:

$$\bar{x} = \frac{1}{n} \sum_{i=1}^{n} x_i = \frac{x_1 + x_2 + x_3 + \cdots + x_n}{n}.$$

In words, the average is simply the sum of the data divided by the number of data points and it is one way to characterize the overall incidence of disease. As shown below, we also need it to calculate the standard deviation of the data.

First, we begin by calculating the *variance*, which is denoted as s^2:

$$s^2 = \frac{1}{n-1} \sum_{i=1}^{n} (x_i - \bar{x})^2$$
$$= \frac{(x_1 - \bar{x})^2 + (x_2 - \bar{x})^2 + (x_3 - \bar{x})^2 + \cdots + (x_n - \bar{x})^2}{n-1}.$$

Now, you may be wondering why we divided the sum of the squared deviations by $n-1$ rather than n. Statisticians use the variance to *estimate* the unobserved variance of the population and it turns out that dividing by $n-1$ makes the estimation better. Statisticians would say that s^2 is an *unbiased* estimator, but these details needn't bother us here.

The standard deviation is the square root of the variance: $s = \sqrt{s^2}$. The important thing to note is that the standard deviation s is a good way to summarize how noisy a set of data is. Larger values of s mean there is more noise, or variability, in the data.

Now, you may be asking: Why do we care about quantifying the noise in data? Here are two reasons. First, it is a nice way to summarize and thus provide some intuition about how

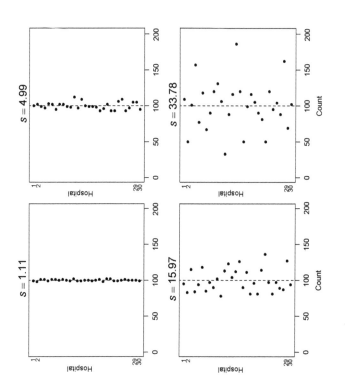

Figure 2.3 Plots illustrating the sample standard deviation for counts of people going to $n = 30$ hospitals with flu-like symptoms. In all of the plots, the average is 100 but the standard deviations range from slightly larger than one to just under 34. The plots show that, as the standard deviation increases, the points are more spread out around the average.

variable a set of data is. Think of it like the margin of error in a poll; the smaller the margin of error the more one might believe the poll results reflect the opinions of the larger population. Second, as we will describe in the next section, we can use the standard deviation to help separate the normal from the unusual.

A Closer Look

Another reason the standard deviation can be useful is something called the *empirical rule*. The empirical rule says that, if the data follows a normal distribution (the classic bell shaped curve), then 68% of the data should fall within one standard deviation of the population average (often called the "mean"), 95% within two standard deviations, and 99.7% within three standard deviations. Thus, if the empirical rule holds, it is highly unlikely that we would get an observation more than three standard deviations away from the mean. There are only three chances in a thousand that this would happen.

So, if we see a data point more than three standard deviations away from the average, there are only three possible explanations: (1) the data do not follow a normal distribution, (2) a very rare thing happened, or (3) perhaps there is something unusual occurring and our assumed normal distribution is not the correct one. Assuming we have already ruled out (1) as a possibility and are willing to accept that (2) is so rare that we don't believe it happened, we can confidently conclude that something unusual occurred. A value that lies outside of the range that we would expect to occur by chance is called an *outlier*.

2.3 LOOKING FOR A SIGNAL OVER TIME

To illustrate how one might look for unusual occurrences over time, say a flu outbreak, let's explore the idea of control charts. The most popularly used control chart is the *Shewhart control chart* named after its inventor, Walter A. Shewhart who developed the method in 1924 while working at Western Electric's manufacturing plant in Cicero, Illinois. The idea is simple: data are taken over time and either the data or summary statistics from the data are plotted, and the plot is used to help identify deviations from what would be expected if the process is in a stable normal state.

For our purposes, let's focus on the *individuals control chart* that is used for monitoring using individual observations. Here the idea is to look for large deviations from a *target value*, which is typically the process mean, where a large deviation is usually defined as an observation outside of the mean plus or minus three standard deviations. As you'll remember, if the process follows a normal distribution we would only expect 0.3% of the observations to fall outside of this range.

Returning to our flu example, imagine we are now monitoring the aggregate number of patients coming into all 30 hospitals with influenza-like illness (ILI) on a daily basis. (We will discuss this further in Chapter 5.) On a normal day, when there isn't a seasonal flu outbreak, we would expect 100 patients on average to come into each hospital each day with ILI. This information, which is based on large amounts of prior data, then becomes an average of $100 \times 30 = 3,000$ across all 30 hospitals. Let's also say that standard deviation of the aggregate number of patients when things are normal is $s = 200$. Then we can establish an *upper control limit* or UCL and an *lower control limit* or LCL on this process as:

$$\text{UCL} = 3,000 + 3 \times 200 = 3,600$$
$$\text{LCL} = 3,000 - 3 \times 200 = 2,400.$$

With this, we can now set up an individuals control chart and monitor the aggregate number of people who come to all 30 hospitals each day. If that number spikes up above the UCL, then we have evidence that the number of people with flu-like symptoms has significantly increased, and perhaps we have a potential flu outbreak. Figure 2.4 shows a control chart for some hypothetical data following our scenario. We simulated the data so that between days 1 and 90 things were normal, but on day 91 and onward a flu outbreak occurred in which the average number of people going to the hospitals

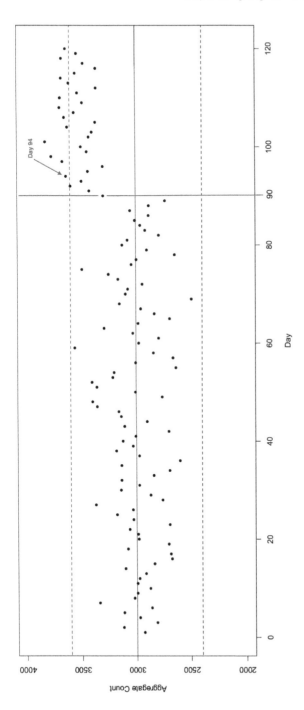

Figure 2.4 A hypothetical control chart for detecting flu outbreaks, where we see that all the points are within the control limits on or before day 90. However, when the outbreak occurs from day 91 onwards, we can see that in general the points are visually shifted upwards, though the chart does not signal until day 94.

with flu-like symptoms increased 15% to an average of 3,450 per day.

What we see in Figure 2.4 is that all the points are within the control limits on or before day 90. Then we can see from day 91 onwards, when the outbreak occurs, the points are visually shifted up, though the first signal, meaning a point that exceeds the upper control limit, does not occur until day 94. What the control chart does is allow public health professionals to ignore the day-to-day variation inherent in the process when things are normal, but to quickly identify when things may no longer be normal.

Note that the control chart didn't indicate a signal until day 94 even though the true process changed at day 91. (We know this because we simulated the data this way. In practice, no one knows when the change actually occurred; we can only go by what the control chart says.) Because of the noise inherent in the system, the control chart will not always immediately raise an alert with a point outside the control limits. In this case, the control chart alerted us on the fourth day of the outbreak. Statisticians have to deal with this uncertainty in trying to separate the signal (a true outbreak) from the noise (the inherent variability).

If we've properly designed the chart, the chance of a "false positive" occurring, meaning the chart signals because a point falls outside the control limits even though the process remains stable, is just three chances out of a thousand. If we'd like to reduce the chance of a false positive signal, we can increase the number of standard deviations around the target level, though that comes at the cost of potentially delaying a true signal once an outbreak occurs. Conversely, we can decrease the number of standard deviations around the target value in order to increase the sensitivity of the chart to detecting outbreaks, but that also comes at the cost of increasing false positive signals. Ultimately, the user must make a decision in setting the control limits that requires a trade-off between speed of detection and the rate of false positives.

2.4   CYCLIC PATTERNS IN DISEASE

For many diseases, the incidence rate varies cyclically throughout the year. Some diseases such as the seasonal flu and pertussis (whooping cough) peak in the winter when the weather forces people to remain indoors, and therefore closer together. Other diseases, including foodborne diseases like *E. coli* and *Salmonella* poisoning, peak in the summer when people often eat outside. Other diseases, such as various forms of hepatitis, are fairly stable across the year. For diseases that are cyclic, the cyclic pattern must be accounted for when determining the upper and lower control limits on the chart.

Pertussis, caused by the bacterium *Bordetella pertussis*, is a respiratory disease that causes a violent cough that is often followed (or preceded) by an inhaling "whoop" sound. It is communicable through droplets coughed out by the patient and often spreads among children. Adults, however, make up about one-fourth of all cases. A vaccine exists for pertussis, and it is usually given as part of the DPT vaccine that children are given.[3] According to the CDC, the vaccine rates for children aged 19 to 35 months is about 85%.

In Missouri, as in most states, pertussis is a reportable disease. This means that physicians must report to the state any cases of pertussis. The Missouri Department of Health and Senior Services monitors all reportable diseases in the state and Figure 2.5 shows the numbers of cases of pertussis in Missouri for every week between 2004 and 2013. There is clearly a cyclical feature to the data, with lower counts in the summer (roughly weeks 24 through 37) and higher in the late fall and winter (roughly week 42 through week 10 of the following year).

Statistically Speaking

The yearly cyclic pattern resembles the sine or cosine functions that students study in trigonometry. These functions form the foundation for modeling the data on disease counts. The model must account for: (1) the overall level (high or low) of the disease, (2) the amplitude (how much higher and lower the disease counts get relative to the overall level), and (3) the phase shift (which describes where in the cycle the process begins).

A function of the form

$$f(t) = \beta_0 + \beta_1 \sin \frac{2\pi t}{52} + \beta_2 \cos \frac{2\pi t}{52}$$

allows us to determine all three of these parameters, assuming the data are reported weekly (52 times per year). Thus, for a given disease, we would take the data, estimate the parameters $\beta_0, \beta_1, \beta_2$, and then use the function above as a model for the expected counts on the control chart. Both the expected level of the disease and the upper and lower control limits are then cyclic.

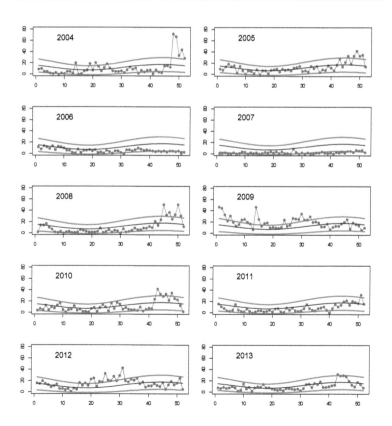

Figure 2.5 Reported cases of pertussis in Missouri by week from 2004 through 2013.

As shown in Figure 2.5, regardless of the time of year, there are always at least *some* cases of pertussis. The blue dots show the actual counts, and the black curve shows the fitted trigonometric curve (see the *Statistically Speaking* box above). The red curves indicate the upper and lower control limits. We can see that there were several outbreaks.

The greatest number of cases occurred in the 2004 outbreak near the end of the year when there were two weeks in succession with 72 and 67 cases. There was also a rather long outbreak that began in late 2008 and didn't subside until the following fall. There were also outbreaks in the fall/winter of 2010 and again in the summer of 2012. For many years, such as 2006, 2007, 2011, and 2013 there were more weeks where the count was below what was expected and there were no signals (with the exception of one signal in week 43 of 2013 where a point was just above the upper control limit).

Notes

[1]Source: www.cdc.gov/flu/about/season/flu-season.htm.

[2]For those who are interested in such details, see Fricker, Jr., R.D. (2013). *Introduction to Statistical Methods for Biosurveillance, with an Emphasis on Syndromic Surveillance*, Cambridge University Press.

[3]The "P" stands for pertussis. The other letters stand for diphtheria and tetanus.

CHAPTER 3

Types of Public Health Surveillance

SEEKING to avoid pandemics, as well as disease outbreaks in general, public health agencies conduct a variety of health-related surveillance activities. The idea is to gather and analyze data related to human health and disease with the hope of providing early warnings of adverse health events and emerging disease outbreaks. Known as disease or *epidemiologic surveillance*, as shown in Figure 3.1, this is just one aspect of a broader set of public health surveillance activities that also encompass the tracking of adverse reactions to medical interventions (particularly drugs and vaccines) and how health and medical services are used. For our purposes, we will just focus here on epidemiologic surveillance.

Epidemiologic surveillance has two main objectives: to enhance outbreak *early event detection*, that is, to identify the outbreak as quickly as possible, and to provide outbreak *situational awareness*, that is, an assessment of the extent of the situation.[1]

Early event detection is critical for catching an outbreak as soon as possible, so that medical and public health personnel can intervene before the outbreak grows into a pandemic. Situational awareness is essential for understanding when and

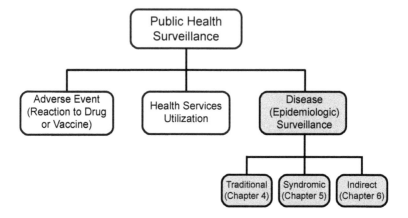

Figure 3.1 Taxonomy of public health surveillance activities. Disease (epidemiologic) surveillance is but one part of a broader set of public health surveillance activities.[2]

where to intervene as well as whether the intervention is having the desired effect. It is focused on monitoring an outbreak's magnitude, geography, rate of change, and life cycle.

Epidemiologic surveillance is one part of *biosurveillance*, which is health surveillance of the entire biosphere – humans, animals, and plants – where as we have already seen in Chapter 1, one of the major pandemic risks is an animal disease that mutates and then is transmitted to and within human populations.

Biosurveillance can be defined as

> the process of active data-gathering with appropriate analysis and interpretation of biosphere data that might relate to disease activity and threats to human or animal health – whether infectious, toxic, metabolic, or otherwise, and regardless of intentional or natural origin – in order to achieve early warning of health threats, early detection of health events, and overall situational awareness of disease activity.[3]

"Biosphere data" are divided into information about human, animal, and agricultural populations, and biosurveillance thus consists of health-related surveillance on each of these populations.

3.1 EPIDEMIOLOGIC SURVEILLANCE

The World Health Organization (WHO) defines epidemiologic surveillance as "the continuous, systematic collection, analysis and interpretation of health-related data needed for the planning, implementation, and evaluation of public health practice."[4] The WHO goes on to say that epidemiologic surveillance can:

- serve as an early warning system for impending public health emergencies;

- document the impact of an intervention, or track progress towards specified goals; and

- monitor and clarify the epidemiology of health problems, to allow priorities to be set and to inform public health policy and strategies.

So, as we have already described, the main idea of epidemiologic surveillance is to collect health data on human populations. This data collection can be either *active* or *passive*. Passive surveillance is based on the regular, ongoing reporting of diseases and conditions by some or all health facilities in a particular region. In contrast, active surveillance involves deploying health professionals to the field to collect data, often for an outbreak that has already been identified. This is often referred to as *event-based* surveillance. In event-based surveillance the investigation is a *retrospective* effort focused on trying to determine the cause of a known outbreak. In contrast, today epidemiological surveillance can also be a *prospective* exercise in monitoring populations for potential disease outbreaks by routinely evaluating data for evidence of an outbreak prior to the existence of a confirmed case and perhaps even prior to any suspicion of an outbreak.

A Closer Look

When a surveillance system detects a potential major outbreak, the CDC's Global Rapid Response Team (RRT) has over 50 highly trained responders ready to deploy on short notice anywhere in the world. In 2016 the Global RRT staff responded to more than 90 outbreaks from cholera to yellow fever, Ebola, measles, polio, and a number of natural disasters.[5]

For example, in 2016 the Global RRT deployed to Haiti in response to Hurricane Matthew. There, in conjunction with the Haiti Ministry of Health (MOH) and CDC Haiti staff, the Global RRT, "focused on assessing storm damage to health care facilities, working with the MOH to rebuild surveillance systems, investigating disease outbreaks, and coordinating with other US and international partners involved in the response. [F]ield work in the first days of the response helped identify an increase in suspected cholera cases related to Hurricane Matthew's widespread destruction of infrastructure, health facilities, and water systems in the southwest portion of the country."[6]

As we'll discuss in more detail in Chapters 4 and 5 many public health organizations, including the WHO and the CDC, have created surveillance systems to identify and then track disease outbreaks. These systems monitor for diseases such as influenza, bird flu, Ebola, and even bioterrorism, such as the 2001 anthrax attacks in the United States that we will explore more in Chapter 11.

3.2   ZOONOTIC SURVEILLANCE

An emerging paradigm that recognizes that humans and animals are interconnected in many ways, including disease, is called *One Health*. In particular, for our purposes, One Health recognizes that human health is connected to animal health and the environment, including *zoonoses* (diseases that normally exist in animals but that can infect humans, much like the bird flu did in the scenario described in Chapter 1).

As a result, more emphasis is being placed on establishing systems for zoonotic surveillance. For example, the Global Early Warning System for Major Animal Diseases, including Zoonoses (GLEWS), was established by the WHO, the Food and Agriculture Organization (FAO) of the United Nations, and the World Organization for Animal Health (OIE). GLEWS is a joint system that builds networks to assist in early warning, prevention, and control of animal disease threats, including zoonoses.

Because Ebola is highly virulent in nonhuman primates and because it can be transmitted to humans, it makes sense to perform zoonotic surveillance on apes and monkeys. Zoonotic surveillance could also be useful for the protection of animal food supplies from disease and wildlife conservation and preservation.[7]

A Closer Look

Not all diseases are communicated by person-to-person contact. In the United States, *Salmonella* is estimated to sicken about 1.2 million people yearly. Within three days after infection, most people develop diarrhea, fever, and abdominal cramps. The illness usually lasts four to seven days, and most people recover without treatment. However, for some the diarrhea may be so severe that hospitalization will be required and, in extreme cases, *Salmonella* can be fatal. It is estimated that it kills 450 people per year in the US.[8]

We often think of *Salmonella* as being caused by contaminated food, but it can also be caused by exposure to animals such as pet turtles. From March 1, 2017 to October 14, 2017, 66 people across 18 states were infected with *Salmonella Agbeni*, where the CDC linked the outbreak to contact with pet turtles and things such as the water from turtle habitats.

Despite appearances of cleanliness and health, all turtles can carry the *Salmonella* bacteria. As a result, and because of the link to *Salmonella* infections, particularly in children, since 1975 the United States Food and Drug Administration (FDA) has banned selling small turtles (defined as turtles with shells less than four inches long) as pets.[9]

3.3 AGRICULTURAL SURVEILLANCE

Agricultural surveillance includes the monitoring of livestock and plant diseases important to the human food chain. As described by the FAO of the United Nations, the spread of plant pests and diseases has increased dramatically in recent years, affecting food crops and causing significant losses to farmers as well as threatening food security. Left unchecked, pests and diseases can spread quickly within a country and even across borders to reach epidemic proportions, where outbreaks can cause massive crop losses that threaten farmers' livelihoods and the food security of millions of people.

As described by the FAO, plant pests and diseases spread in three principal ways:

- human-based movement related to trade and travel,

- environmental forces such as wind and weather, and

- insect and other animal hosts.[10]

Think About It

Fusarium wilt of banana (FW), also known as Panama disease, is a lethal plant disease caused by a soil-borne fungus. As reported by the FAO, it is one of the most destructive banana diseases worldwide and has been causing serious losses in Southeast Asia. As of 2017, it has spread to the Middle East, Mozambique, and South Asia and is likely to continue spreading.

Bananas and plantains are the most exported fruit in the world, where some 400 million people rely on them as a food staple and/or a source of income. FW can cause 100 percent crop loss and, thus far, it cannot be controlled using fungicides nor can it be eradicated from soil using fumigants. Once the fungus is established in a location, continued production of bananas in infested soils relies on replacing susceptible banana varieties with resistant ones.

The United States Department of Agriculture (USDA) operates the National Animal Health Surveillance System (NAHSS) with the goal of systematically collecting, compiling, and analyzing animal health data for the purpose of protecting animal health and the food supply. Similarly, the Department of Agriculture and Food, Western Australia (DAFWA) conducts disease surveillance and testing programs for nationally important diseases such as mad cow disease and bluetongue virus.

Notes

[1] The United States Centers for Disease Control and Prevention (often referred to as the CDC) defines them this way. Early event detection is the ability to detect, at the earliest possible time, events that may signal a public health emergency, and situational awareness is the ability to utilize detailed, real-time health data to confirm or refute, and to provide an effective response to, the existence of an outbreak.

[2] Source: Fricker, Jr., R.D. (2013). *Introduction to Statistical Methods for Biosurveillance, with an Emphasis on Syndromic Surveillance*, Cambridge University Press, p. 6.

[3] Homeland Security Presidential Directive 21 (HSPD-21).

[4] Source: `www.who.int/topics/public_health_surveillance/en/`.

[5] Source: `www.cdc.gov/globalhealth/healthprotection/errb/global-rrt.htm`.

[6] "Global Rapid Responders – Our Boots on the Ground Defense" by Ashley Greiner, dated February 15, 2017: `https://blogs.cdc.gov/global/2017/02/15/global-rapid-responders-our-boots-on-the-ground-defense/`.

[7] See `www.ncbi.nlm.nih.gov/books/NBK207997/`.

[8] See `www.cdc.gov/salmonella/general/technical.html`.

[9] For more information, see `www.cdc.gov/salmonella/agbeni-08-17`.

[10] See `www.fao.org/emergencies/emergency-types/plant-pests-and-diseases/en/`.

CHAPTER 4

Traditional Surveillance

PUBLIC health organizations gather data from a number of sources for their traditional surveillance activities. These include various surveys of populations and, most notably, reporting by hospitals and physician's offices. Indeed, there are a number of serious diseases and conditions that hospitals and doctors are required to report to state and local public health authorities. As we briefly explored in Chapter 3, public health surveillance focuses on collecting data on adverse events related to drugs and vaccines, the utilization of various types of health-related services, and disease surveillance. Here we explore some of the traditional methods used for disease surveillance, and then in Chapter 5 we'll learn a bit about syndromic surveillance. We'll then look at other approaches to disease surveillance, such as monitoring social media, in Chapter 6.

4.1   SUMMARIZING DATA

The Centers for Disease Control and Prevention's *Morbidity and Mortality Weekly Report* is a series of reports published by the CDC. They are used for the "scientific publication of timely, reliable, authoritative, accurate, objective, and useful public health information and recommendations."[1] The MMWR's readership includes physicians, nurses, public health practitioners, and epidemiologists.

In particular, the MMWR publishes *Surveillance Summaries* to disseminate a variety of surveillance findings with detailed interpretations of trends and observed patterns. One such disease is Lyme disease, which is caused by the bacterium *Borrelia burgdorferi*. Symptoms are much like those for the flu (nausea, muscle pain, chills, and fever) but they can last for a long time. The most noticeable and identifiable symptom is a "bull's eye" rash emanating from an insect bite. If caught early, it can be treated with antibiotics; if it is not caught early, Lyme disease can become chronic. Most chronic Lyme patients exhibit symptoms such as fatigue, sleep impairment, joint pain, muscle pain, cognitive impairment, neuropathy, and others[2]. Patients often experience a lower quality of life than many other chronic diseases, such as multiple sclerosis, congestive heart failure, and diabetes.[3]

In November 2017 MMWR published "Surveillance for Lyme Disease – United States, 2008-2015,"[4] that summarized Lyme disease cases reported to the CDC via the National Notifiable Diseases Surveillance System (NNDSS). The way the system works is that doctors and testing laboratories report possible cases to local and state health departments which then investigate the reports and classify them according to the national surveillance case definition. Those that qualify as confirmed or probable cases of Lyme disease are then reported to CDC through NNDSS.

Figure 4.1 shows the incidence of the disease over time, where between 2008 and 2015 a total of 275,589 cases of Lyme disease were reported to CDC (208,834 confirmed and 66,755 probable). Nationwide, the number of reported cases increased from about 10,000 per year in 1992 to nearly 40,000 within twenty years. Figure 4.2 shows that most cases were reported from the Northeast, mid-Atlantic, and upper Midwest regions. The disease has spread geographically, although the report concludes that "case counts in most of these states have

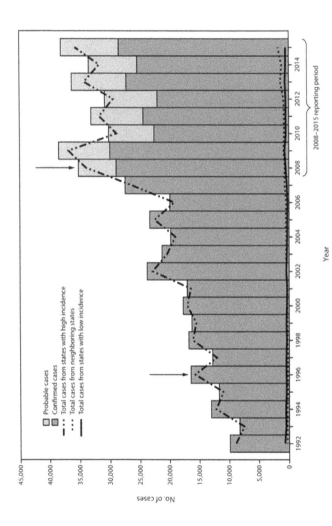

Figure 4.1 From "Surveillance for Lyme Disease – United States, 2008-2015" showing the number of confirmed and probable Lyme disease cases by year, from 1992 to 2015, in the United States.[4]

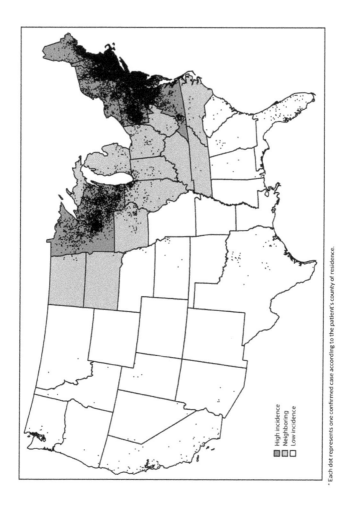

Figure 4.2 From "Surveillance for Lyme Disease – United States, 2008-2015" showing the average annual number of confirmed Lyme disease cases by county of residence in the United States.[4]

remained stable or decreased during the reporting period." Lyme disease has now been reported in all 50 states. Statisticians would refer to these statistics and associated plots as *descriptive statistics*. They describe the incidence of the disease both temporally and spatially and, in so doing, help inform medical and public health practitioners about the current situation of Lyme disease. Returning to the objectives of disease surveillance first discussed in Chapter 3, this type of information is useful for enhancing Lyme disease situational awareness.

4.2 THE HISTORICAL LIMITS METHOD

The CDC's National Notifiable Diseases Surveillance System is one example of event-based surveillance. As we've just seen, it aggregates and summarizes data on certain diseases that health care providers are required by law to report to public health departments. Here we use the NNDSS to illustrate the *historical limits* method for disease outbreak detection.

The historical limits method compares the observed incidence of a particular disease from a current time period to incidence data from equivalent historical periods. For example, as shown in Figure 4.3, a plot from the MMWR's "Notifiable Diseases and Mortality Tables," the NNDSS uses it to characterize the current incidence of various reportable infectious and noninfectious diseases, such as anthrax, cholera, plague, polio, and smallpox, where each week every state reports cases counts for each of the reportable diseases to the CDC, and then the CDC compares the counts for each disease to historical average incidence rates from equivalent periods in the past five years.

Statistically Speaking

To do the calculations for the historical limits method, let $T_{i,j,k}$ be the most recent 4-week total for reportable disease i, in week j and year k. This is then compared to the average total

number of cases reported for the same 4-week period, the preceding 4-week period, and the succeeding 4-week period for
the previous five years as follows. First, the average of the 15
historical 4-week periods is calculated as

$$\bar{x}_{i,j,k} = \frac{1}{15} \sum_{s=1}^{5} \sum_{r=-1}^{1} T_{i,j-r,k-s}.$$

Then the standard deviation is calculated as

$$s_{i,j,k} = \sqrt{\frac{1}{14} \sum_{s=1}^{5} \sum_{r=-1}^{1} (T_{i,j-r,k-s} - \bar{x}_{i,j,k})^2},$$

where the historical upper and lower limits (UL and LL) for
reportable disease i, in week j and year k, are

$$UL_{i,j,k} = \bar{x}_{i,j,k} + 2s_{i,j,k} \text{ and } LL_{i,j,k} = \bar{x}_{i,j,k} - 2s_{i,j,k}.$$

This is very much like the control chart idea from the previous
chapter, except the limits are based on two standard deviation
differences from the average rather than three standard deviation differences, and only data from comparable previous years'
time periods are used to estimate the mean and standard deviation.

Finally, the plot in Figure 4.3 is based on the idea of "standardiz-
ing" the totals by the historical average, $T_{i,j,k}/\bar{x}_{i,j,k}$, and plotting on a log scale. So, if the current total is the same as the
historical average total, then the ratio is 1 in the plot, and the
limits are

$$UL^*_{i,j,k} = 1 + 2s_{i,j,k}/\bar{x}_{i,j,k} \text{ and } LL^*_{i,j,k} = 1 - 2s_{i,j,k}/\bar{x}_{i,j,k}.$$

Plotting the ratio allows for a direct comparison between the observed total and the historical average, but it also means each
disease is on a different scale and so the cross-hatching is required to show when and by how much a disease's total count
exceeds the upper or lower limit.

Figure 4.3 shows the log ratio of the observed count to the
historical average, and it tells when the counts observed in the
past four weeks exceed the average observed in the past five
years plus or minus two standard deviations. For example, the

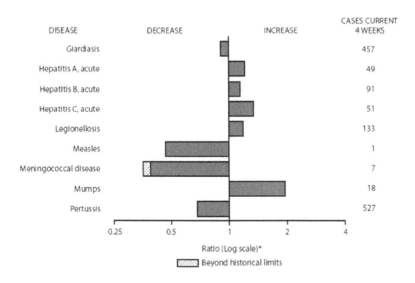

Figure 4.3 From "Notifiable Diseases and Mortality Tables" for week 52 of 2015. For this week, the Meningococcal disease count was below its lower historical limit.[5]

gray cross-hatching on the bar associated with Meningococcal disease in Figure 4.3 means that the number of cases of meningococcal disease reported in the last week of 2015 was more than two standard deviations *below* its historical norm. This is good news, where it means we have solid evidence of a decrease in the incidence of meningococcal disease which is caused by the type of bacteria called *Neisseria meningitidis*. This disease includes infections of the brain and spinal cord (meningitis) and bloodstream infections (bacteremia or septicemia) that are often severe and can be fatal.

On the other hand, all of the other diseases shown in Figure 4.3 (giardiasis; acute hepatitis A, B, and C; legionellosis; measles; mumps; and pertussis), while deviating to a greater or lesser degree from their historical rates, are not statistically different. That is, for these diseases we do not have sufficient evidence to conclude there the observed deviations are anything more than the amount of variation that would normally be expected. More simply put, it's reasonable to conclude that the fluctuations in these disease counts are just noise.

4.3 TRACKING DISEASES OVER SPACE AND TIME

Surveillance data often have a geographic component to them. For example, we might be tracking pertussis not only across time but also in space, meaning geographically, because we would like to know *where* it occurred in addition to knowing *when* it occurred.

Typically, data are collected across predefined regions, such as states, counties, zip codes, or census tracts. Part of the problem in dealing with such data is the irregular spacing and shape of the regions. Another problem is the vastly different sizes of the regions. For example, in Missouri, St. Louis County has slightly over one million residents while neighboring Franklin County has about one-tenth that amount (just over 100,000). This is why monitoring disease *rates* is important, although the variability is often related to size, so there will be less variability in estimates of disease rates in large counties than in small counties. In a rigorous analysis, this must also be considered.

It is often the case that nearby regions are more correlated than regions that are separated by great distances. For example, the disease rates in St. Louis County and St. Charles County, which are adjacent, are likely to be similar, but Jasper County in the far southwest corner of the state, which contains Joplin, will probably be less correlated. In other words, even when an outbreak is not occurring, a relatively high disease rate in St. Louis County will probably be associated with a high rate in St. Charles County, and vice versa. In contrast, a high disease rate in St. Louis County will say very little about the rate in Jasper county.

For a disease that is rare, the observed disease rates may vary considerably, even in neighboring counties. For example, rates in adjacent counties may be 8, 0, and 3 (per 100,000 people). Because we expect that neighboring regions should have correlated rates, we might think it reasonable that 8 per 100,000 is too high, and 0 per 100,000 is definitely too low as an estimate of the true disease rate. To address this type of situation, statisticians often find it helpful to *smooth* the disease rates based on the correlation.

Sudden Infant Death Syndrome In North Carolina

Consider the cases of sudden infant death syndrome (SIDS) in North Carolina from 1979 to 1983.[6] The data consist of the number of live births and the number of SIDS cases in each of the 100 North Carolina counties between July 1, 1979 and June 30, 1984. The raw SIDS rates by county are shown in Figure 4.4. This is an example of what is called a *choropleth map*, which is a map where the regions, counties in this case, are color coded according to value of some variable. Here the colors represent the rate of SIDS (cases divided by the number of live births), where the darker regions have higher SIDS rates. Note in the choropleth map how the rates in some adjacent regions can be vastly different. If it's reasonable to assume that the rates in nearby counties are similar, and it often is,

Figure 4.4 Raw rates (per 1,000 births) of SIDS in North Carolina.

then these rate differences in neighboring counties is evidence of noisy data.

Note that there are counties whose rates differ substantially from their neighbors. This is especially true in the western, less populated region. (Smaller populations lead to greater variability in rates.) Polk County, for example, has a much lower rate than its neighbors, and Alleghany has a much higher rate than its neighbors. We would expect the rates to be closer together for regions that are near to each other.

In cases like these it is often desirable to *smooth* raw SIDS rates. One method of doing so is to assume a structure where regions that are adjacent are likely to be similar to each other and thus each county is adjusted according to the values observed in its neighbors. For our purposes, we can omit details of the statistical methods used, which can be quite complicated, and just note that if two nearby counties have high raw rates (i.e., unsmoothed or unadjusted) and another county between them has a very low rate, then the rates are brought closer together. The highest raw rates of SIDS are pulled lower, and the lowest rates are pulled higher. The smoothed rates from one method (called a conditional autoregressive model[7]) are shown in Figure 4.5. What we see in this map is that the

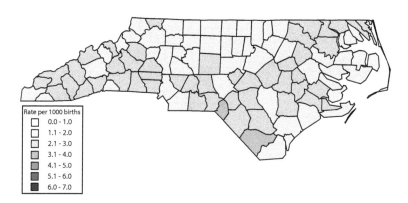

Rate per 1000 births
☐ 0.0 - 1.0
☐ 1.1 - 2.0
☐ 2.1 - 3.0
☐ 3.1 - 4.0
■ 4.1 - 5.0
■ 5.1 - 6.0
■ 6.0 - 7.0

Figure 4.5 Smoothed rates (per 1,000 births) of SIDS in North Carolina using a conditional autoregressive model.

Figure 4.6 Choropleth maps for Connecticut showing the incidence rate of Lyme disease from 1993 to 2016.

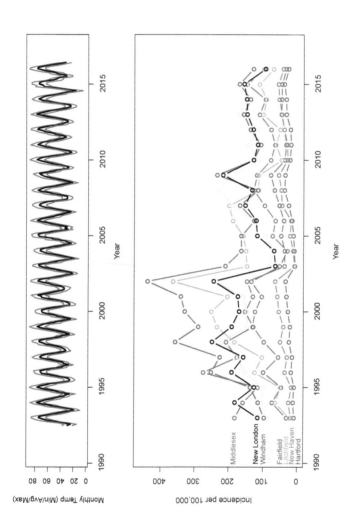

Figure 4.7 Lyme disease in Connecticut's eight counties from 1993–2016. The temperatures are plotted at the top. Note the drop in Lyme disease after the cold winter of 2002/2003.

estimated rates, after smoothing, now vary much more gradually with neighboring counties having more similar rates, as we would expect.

Lyme Disease in Connecticut

Multiple choropleth maps can be used to show the change in disease rates across time and across space. Figure 4.6, for example, shows cases of Lyme disease in Connecticut from 1993 to 2016 and consists of 24 choropleth maps, one for each year. Notice the gradual but steady darkening of the colors in the maps from 1993 until 2002, and then the precipitous drop in 2003.

The 2003 drop can also be seen in a *time series* plot. Figure 4.7 shows the rates in each county from 1993 through 2016. Windham County, a mostly rural county in the northeastern corner of the state, often has the highest rate. This is not surprising since Lyme disease is carried by ticks which are more prevalent in the country than in cities. Also note the plot of monthly temperatures at the top of Figure 4.7. The winter of 2001/2002 was warmer than usual, and in 2002 the Lyme disease rates peaked. The following winter was somewhat colder than most previous winters and the incidence of Lyme disease dropped substantially in 2003. This is probably due to a drop in the tick population after the cold winter.

Notes

[1]Source: www.cdc.gov/mmwr/about.html.

[2]//www.lymedisease.org/lyme-basics/lyme-disease/chronic-lyme-disease.

[3] Johnson, L., Wilcox, S., Mankoff, J., and Stricker, R. B. (2014). Severity of chronic Lyme disease compared to other chronic conditions: a quality of life survey. *PeerJ*, **2**, e322.

[4] To read the complete report, see `www.cdc.gov/mmwr/volumes/66 /ss/ss6622a1.htm`.

[5] Source: `www.cdc.gov/mmwr/preview/mmwrhtml/mm6452md.htm?s_c id=mm6452md_w`.

[6] The SIDS data set was first discussed in Symons, M.J., Grimson, R.C., and Y.C. Yuan (1983). Clustering of Rare Events, *Biometrics*, 39(1), 193–205.

[7] See Lawson, A. (2018). *Bayesian Disease Mapping: Hierarchical Modeling in Spatial Epidemiology*, 3e, CRC Press, Boca Raton.

CHAPTER 5

Syndromic Surveillance

SUPPOSE you wake up one day with a sore throat and generally feel poorly. Thinking it is just the common cold you slog through the day, and the next. By the third day you have muscle aches, joint pain, a cough, and seem to be running a fever, so you make an appointment with your physician. It is the latter part of January and influenza (the "flu") has been particularly high this winter. When you get to see the physician at the end of day three, you describe the symptoms. There are tests for influenza, the simplest being the rapid influenza diagnostic tests (RIDTs). These tests involve a swab from your nose or throat, and they look for antigens that your body would produce to fight the flu. While the results can be available within ten to fifteen minutes, the test is not very accurate. Other more accurate tests are available, but these take longer and are more expensive. Does your physician order the RIDT? Or some other test? Maybe, but maybe not. It is likely that your physician has seen many other cases similar to yours earlier in the day, and believes with reasonable confidence that you indeed have the flu. To save time and expense, your physician may forego the test. The main reason to skip the test may be that your treatment would be same regardless of whether the test is positive or negative. If no influenza test is performed, your case would be categorized as an *influenza like illness* (ILI), not a confirmed case of influenza. This is a subtle, but important, difference.

Syndromic surveillance is a specific type of epidemiologic surveillance that is based on *indicators* of diseases and outbreaks, not confirmed cases, with the goal of detecting outbreaks before medical and public health personnel would otherwise notice them. Syndromic surveillance differs from traditional epidemiologic surveillance in a number of important ways. First, it often uses non-specific health and health-related data (e.g., daily number of individuals reporting to an emergency room with sore throats) whereas traditional notifiable disease reporting is based on suspected or confirmed cases (e.g., daily number of individuals diagnosed with the meningococcal disease). Second, traditional public health surveillance is generally conducted on specific, well-defined diseases, and it is generally not initiated without a known or suspected outbreak.

In contrast, syndromic surveillance actively searches for evidence of possible outbreaks, perhaps even before there is any suspicion that an outbreak has occurred. The motivating idea is that more serious diseases often first manifest with more benign symptoms before they can be diagnosed for what they really are. For example, for the first week or two after someone contracts smallpox, he or she exhibits flu-like symptoms. Thus, a widespread smallpox outbreak might first become evident as an increase in the counts of the syndrome "influenza-like illness" (ILI) counts before the first smallpox case is diagnosed by a clinician.

5.1 WHAT IS A SYNDROME?

Syndromic surveillance is based on the notion of a *syndrome*, which is a set of non-specific pre-diagnosis medical and other information that may indicate the presence of a disease. Table 5.1 lists some typical syndromes used in syndromic surveillance systems. Compared to traditional disease

Table 5.1 Typical syndromes used in syndromic surveillance ystems.

Botulism-like	Lymphadenitis
Fever	Neurological
Gastrointestinal (GI)	Rash
Hemorrhagic Illness	Respiratory
Influenza-like Illness (ILI)	Unspecified infection

Table 5.2 Chief complaint examples that could be used in a syndromic surveillance system.

```
NEWPAT EST CARE/NO MEDS/ NO CONDITION
OBCPE/RES PER PT SHE WANTS SOONER APPT
DISCHG/MUDGE REV/JP
CHDP /RV PER MOM RES
NEW PT ER FU PT HAD SAB/LM
2WK OB CK PER DR L
FEVER, SORETHROAT, DIARRHEA/LC
NEW/EVAL/SEISURES
NPE FU HOSPITALIZATION DETOX
W/I BCM / ABDL BUMP
BOOK PER MD HTN ///R/S FROM 07/09-NA LFT MESS ON V
C/O STOOL PROB X 1WK//MS
WANTS REFILL/ANY PROV AVAIL
WALK IN FEVER W/RESIDENT
FEELING WEAK,DIZZY-OD
RTN 2WK NB EVAL-VS
NPE WCC CCAH
```

surveillance methods and approaches, which tend to focus on actual diagnoses and diagnostic laboratory results, syndromic surveillance uses the least medically specific data. Often syndromic surveillance is based on data derived from "chief complaints" of people who go to hospital emergency rooms (ERs).

A chief complaint is a medically terse description of the main reason or reasons a person goes to the ER. Table 5.2

gives some examples of actual chief complaints as taken from a syndromic surveillance system. Written by medical personnel, chief complaints contain medical jargon, acronyms, and abbreviations for use by other medical professionals. To distill the chief complaints down into syndrome indicators, the text is searched and parsed for key words, often of necessity including all the ways a particular key word can be misspelled, abridged, and otherwise abbreviated. For example, to generate respiratory syndrome counts, the chief complaint text would be searched for terms such as "respiratory" (including common misspellings such as "resparatory," common typos such as "repsiratory," and common abbreviations such as "resp") and related terms such as "apnea," "bronchitis," "pneumonia," and their common misspellings, typos, and abbreviations.

Figures 5.1 and 5.2 illustrate what syndromic data looks like. This data set was derived from patient-level chief

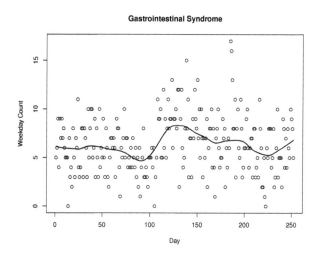

Figure 5.1 One year of gastrointestinal (GI) syndrome data. Each circle is the daily count of the number of people who were classified into the gastrointestinal syndrome. The smooth black curve is a smoothed estimate of the mean response based on local data (i.e., data taken at nearby time periods); this is called a LOESS curve.

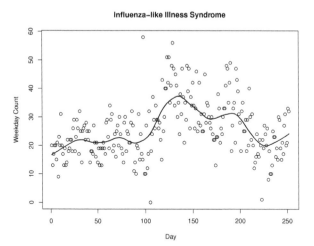

Figure 5.2 One year of ILI syndrome data from the clinics. Each circle is the daily count of the number of people presenting who were classified into the influenza-like illness syndrome. The smooth black curve is the LOESS curve.

complaints for six public health clinics located in the same US county for one year from August 1st through July 31st. Figures 5.1 and 5.2 plot the daily counts for gastrointestinal and influenza-like illness syndromes, where the circles are the daily counts and the line is the estimated daily average count. Note that the clinics were not open on weekends and major holidays, resulting in 252 days with syndrome counts.

For the gastrointestinal syndrome, Figure 5.1 shows that the daily counts range from 0 on three days to a maximum of 17 that occurred on April 28th (day number 185). For the year, an average of 6.4 people with GI symptoms came to the clinics each day. The line shows the daily counts were essentially constant from August through December (days 1-100 or so), followed by an increase in counts from mid-January to mid-March (roughly days 110-150 or so) and a spike in late April (roughly days 185-190), after which the count essentially returned back to the average. This is a pattern consistent with a winter flu season and the ILI syndrome exhibits a similar pattern, where the daily counts range from zero on one day to a maximum of 58 that occurred on December 22 (day 98). Over the entire period, there is an average of 28.8 people with ILI symptoms per day. The line shows a more pronounced increase in ILI in the winter months as well as a spike again in late April.

5.2   SYNDROMIC SURVEILLANCE SYSTEMS

A syndromic surveillance system has four main functions: data collection, data management, analysis, and reporting. As illustrated in Figure 5.3, raw data enters the system at the left and as it flows through the system becomes actionable public health information at the right. Note that these steps are essentially the same as those in a traditional surveillance

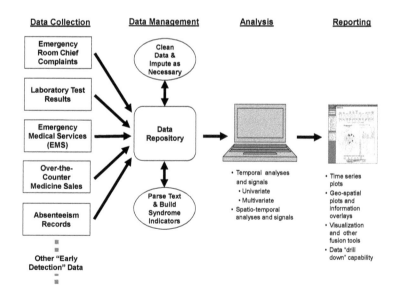

Figure 5.3   An illustrative syndromic surveillance system with its four main functions: data collection, data management, analysis, and reporting.[1]

system; the main difference is in the data set that is ingested in the system. For traditional surveillance systems, the data are suspected or confirmed cases of specific diseases. For syndromic surveillance, it is often the chief complaints that we just discussed. For other types of biosurveillance systems it could be a variety of non-specific but medically relevant data such as laboratory test results, ambulance (EMS) calls, over-the-counter medicine sales, and even absenteeism records, as well as zoonotic and agriculturally related information.

BioSense

Developed and operated by the Centers for Disease Control and Prevention, the CDC's National Syndromic Surveillance Program (NSSP) BioSense platform is intended to be a United States-wide syndromic surveillance system. Begun in 2003, BioSense initially used Department of Defense and Department of Veterans Affairs outpatient data along with medical laboratory test results from a nationwide commercial laboratory and in 2006 it began incorporating data from civilian hospitals as well. The primary objective of BioSense is to "expedite event recognition and response coordination among federal, state, and local public health and healthcare organizations." BioSense has undergone a number of evolutionary changes over the years, including the incorporation of ESSENCE (Electronic Surveillance System for the Early Notification of Community-based Epidemics) within the BioSense platform.

ESSENCE

ESSENCE was developed by Johns Hopkins University for the Department of Defense in 1999 and ESSENCE IV now monitors for infectious disease outbreaks at more than 300 military treatment facilities worldwide on a daily basis using data from patient visits to the facilities and pharmacy data. For the Washington, DC area, ESSENCE II monitors military and civilian outpatient visit data as well as over-the-counter pharmacy sales and school absenteeism. Components of ESSENCE have been adapted and used by some local and state public health departments and, as just mentioned, is now incorporated into BioSense.

Statistically Speaking

The design of an outbreak signaling algorithm proceeds in two steps. First, one must model the "normal" state of disease incidence in the population, which will naturally fluctuate, and which for the purposes of detecting outbreaks is just noise. Then, second, one must monitor deviations from that normal background incidence to look for outbreaks, part of which requires setting some sort of threshold level above which the algorithm will produce a signal.

Without going into all of the details here, there are any number of ways to model the normal disease incidence in the population. For our purposes, let Y_i denote the observed ILI syndrome count on day i and let $\hat{Y}_i$ denote the predicted count from some model that could be based on prior days' counts, population information, and perhaps other parameters to account for, say, seasonality. What we're interested in monitoring are the residuals from the model, $r_i = Y_i - \hat{Y}_i$, where perhaps if the residuals for a sequence of time periods are unusually large, that might indicate an outbreak (since the observed ILI syndrome counts are larger than what we would have expected to see).

Borrowing from the statistical process monitoring literature, one algorithm useful for assessing whether an outbreak is occurring is the cumulative sum or CUSUM. It works by appropriately summing up observations (in this case, standardized model residuals, $z_i = r_i/s_{r_i}$, where we will skip the details of calculating s_{r_i} here) and producing a signal when the sum exceeds some threshold. In particular, let the CUSUM at time $i - 1$ be denoted as C_{i-1}, then the CUSUM at time i is calculated as

$$C_i = \max(0, C_{i-1} + z_i - 0.5).$$

Note how the CUSUM works: whenever $z_i > 0.5$, a positive value is added to the cumulative sum at time i. On the other hand, if $z_i < 0.5$ then a negative value is added to the sum, but note that the sum cannot be negative. If a sequence of z values is consistently greater than 0.5, then the cumulative sum will rapidly get large. A signal is generated at time t when $C_t > h$, for some chosen threshold h.

5.3 EXAMPLE: H1N1 IN MONTEREY, CA

To illustrate syndromic surveillance, we'll look at how using a statistical algorithm to track ILI syndrome counts compares to diagnosed case counts for the seasonal flu that occurred in Monterey, California in early 2009 and then two subsequent H1N1 "swine flu" outbreaks.

Figure 5.4 shows those three outbreaks in gray with the weekly percentage of patients *classified* with ILI from the California Sentinel Provider System, the Monterey County

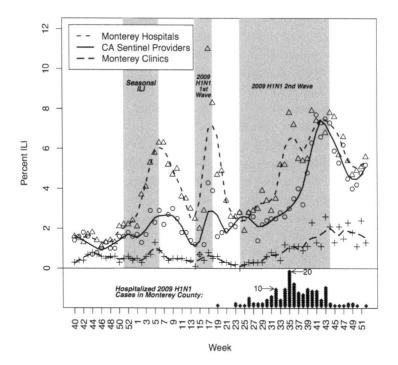

Figure 5.4 Percentage of patients classified with ILI in Monterey County from the week starting on Sunday, September 28, 2008 (week 40) to the week starting on Sunday, December 27, 2009 (week 52). The diamonds are laboratory-confirmed, hospitalized H1N1 cases.[2]

hospital emergency rooms, and the Monterey County public health clinics, along with lines to better show the underlying trend. At the bottom of the figure, the diamonds are laboratory-confirmed, hospitalized cases of 2009 H1N1 in Monterey County, where each diamond represents one person and is plotted for the week the individual first became symptomatic due to 2009 H1N1 infection, which visually also correspond to the second H1N1 outbreak period.

During this same period, the Monterey County Health Department (MCHD) also conducted syndromic surveillance – by monitoring the chief complaint data from the county's four hospital emergency rooms (ERs) and six public health clinics – as an alert system for various types of disease outbreaks, including those naturally occurring (e.g., influenza), accidentally occurring (e.g., fire-related illnesses), or intentionally occurring (e.g., bioterrorism). The MCHD monitored a number of syndromes including ILI, gastrointestinal, upper respiratory, lower respiratory, and neurological.

To distill the chief complaints down into syndrome indicators, the chief complaint text was searched and parsed for key words, where the ILI syndrome was defined as the following text combinations: ("fever" and "cough") or ("fever" and "sore throat") or ("flu" and not "shot"). Thus, if someone went to a Monterey County emergency room and said either that they had a fever and cough, or they said they had a fever and sore throat, or they said they have the flu and did not say they wanted a flu shot, then that person was classified as having the ILI syndrome.

Figure 5.5 is a time series plot of the resulting ILI syndrome counts. The circles are the aggregate daily ILI syndrome counts for Monterey County clinics and the black line is a smoothing line to show the underlying trend. Here we see that the syndrome count trends match quite well with the three outbreak periods, with the black line clearly increasing in each of the three periods suggesting that mining the chief complaint text for appropriate key words does provide information about disease outbreaks.

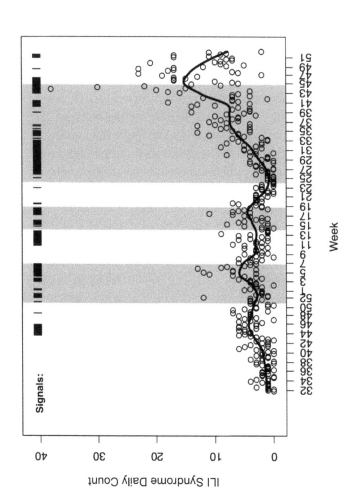

Figure 5.5 ILI syndrome counts over time in Monterey County. The circles are the aggregate daily ILI counts for Monterey County clinics and the black line is a smoothed line to show the underlying trends.[3]

Now, short of looking at this type of plot and subjectively deciding whether an outbreak is occurring, a key part of early event detection is implementing a statistical algorithm that will appropriately signal to public health practitioners that an outbreak might be occurring. The results of one such algorithm is shown at the top of Figure 5.5, where a signal on a particular day is denoted by a vertical line "|" and the heavier black bars indicate a sequence of daily signals. Here we see that the algorithm routinely signals during the outbreak periods as well as sometimes outside of those periods. However, comparing the signals to the time series, we do see that the signals correspond well to increases in the daily ILI syndrome counts, where it could be that the syndromic surveillance may be picking up additional information that was not present or less obvious in the diagnosed ILI time series of Figure 5.4.

Such a signal, like all statistical signals, could either be a true or false positive (just like the lack of a signal could be a true or false negative), and so the signals produced by the algorithm will subsequently require investigation to determine whether it is a true positive, meaning an outbreak is occurring. For readers who are interested, the statistical detection algorithm is briefly described in the "Statistically Speaking" section in this chapter.

Notes

[1] Adapted from Fricker, Jr., R.D. (2013). *Introduction to Statistical Methods for Biosurveillance, with an Emphasis on Syndromic Surveillance*, Cambridge University Press, p. 11.

[2] Source: Hagen, K.S., R.D. Fricker, Jr., K. Hanni, S. Barnes, and K. Michie (2011). Assessing the Early Aberration Reporting System's

Ability to Locally Detect the 2009 Influenza Pandemic, Statistics, *Politics, and Policy*, **2**, issue 1, article 1.

[3]Source: Modified from Hagen, K.S., R.D. Fricker, Jr., K. Hanni, S. Barnes, and K. Michie (2011). Assessing the Early Aberration Reporting System's Ability to Locally Detect the 2009 Influenza Pandemic, Statistics, *Politics, and Policy*, **2**, issue 1, article 1.

CHAPTER 6

Indirect Approaches

G ETTING direct counts on persons who have a confirmed
case of a disease, or even a suspected case, is often difficult. Not everyone who has the flu will seek medical help.
Their first inclination may be to tell family or friends, and
to search for remedies. Even if a person does eventually see
a physician about a possible disease, there is an inherent lag
in the time between the person first showing symptoms and
when the doctor reports the condition.

There are various ways to collect information on diseases,
besides the gold standard of confirmed cases, including internet searches, purchases of items related to a disease (e.g.,
cold/flu remedies, tissues), sick days at work or school, social
media posts, news media, the moderated expert forum, and
futures markets. All of these are indirect ways to obtain information about disease prevalence and we will cover several of
these in the next few sections.

As in syndromic surveillance, these methods yield variables that are *correlated* with the actual disease reports and
can sometimes signal an outbreak before the official reports
are even available. The classical trade-off is between timeli-
ness and validity. The methods described in this chapter give
timely, almost immediate, feedback about the health of a region. Being indirect measures, they can suffer from not measuring what they purport to measure.

6.1   GOOGLE FLU TRENDS

Google's search engine is by far the most popular way to search the internet. The number of unique visitors is four times that of the second most popular search engine and Google has been able to amass huge amounts of data on almost any topic it wants by keeping track of user searches.[1]

Given this, if Google could track search terms like "flu," "influenza," "fever," "muscle aches," and so forth, and if Google could geolocate them (at least to the state or country), then potentially they could estimate the flu rate in nearly real time. Furthermore, since web searches probably precede physician visits by at least several days, monitoring searches related to "flu" might be able to catch an outbreak of influenza several days before a method that relies on physician diagnoses.

With this in mind, Google researchers began studying the relationships between search terms and the percentage of patients with influenza-like illness (ILI) reported by physicians to the CDC. They began with 50 million of the most common search terms and looked at how well each correlated with the CDC ILI data. Using the search term data, from 2007 to 2008 they tested 450 million statistical models to find the one that fit the CDC data best. They found that adding terms yielded stronger correlations with the CDC data up to 45 search terms, after which their correlations got weaker.

In 2008, they launched Google Flu Trends (GFT) and it purported to be able to predict flu rates two weeks faster than the CDC. The actual search terms used in their model have never been revealed by Google, although the general categories were given in a *Nature*[2] paper; see Table 6.1.

Google scientists claimed that

> *Because the relative frequency of certain queries is highly correlated with the percentage of physician visits in which a patient presents with influenza-like symptoms, we can accurately estimate the current level of weekly influenza activity in each region of the United States, with a reporting lag of about one day.*[3]

Table 6.1 Categories of search terms for Google Flu Trends.

Topic	Number Among Top 45
Influenza complications	11
Cold/flu remedy	8
General influenza symptoms	5
Term for influenza	4
Specific influenza symptom	4
Symptoms of an influenza complication	4
Antibiotic medication	3
General influenza remedies	2
Symptoms of a related disease	2
Total	45

This bold statement is an example of what Lazer, Kennedy, King, and Vespignani[4] call *big data hubris*, which "is the often implicit assumption that big data are a substitute for, rather than a supplement to, traditional data collection and analysis."[5] There are several important aspects of this quote, as described in a *Harvard Business Review* paper by Kaiser Fung.[6] He described the OCCAM framework, an acronym for five descriptors of big data.

- Observational. Big data usually come from automatically collected systems, not from some designed experiment to collect data.

- Controls, or really the lack thereof. Without a control group against which to compare, fair and valid comparisons are difficult to make.

- Complete, or seemingly so. Big data can often claim that all outcomes have been captured. For example, data on a hospital system will include *all* patients in the system in a specified time period. Since Google has the lion's share of the search engine market, it's almost as if they are capturing all internet searches.

- Adapted. Usually big data sets were not collected specifically for the purpose at hand. Rather data were collected automatically with little or no plan for the ultimate purpose.

- Merged. Many data sets are merged, or combined together usually ignoring precise definitions of variables and the differing objectives of the original data sets.

Statistically Speaking

Big Data Analytics.   When the amount of data is large, researchers are often looking to make predictions, not inferences. For example, if a store is looking at your propensity to buy swim wear, they will keep track of your purchases of everything and try to see what kinds of items swimmers purchase besides swimming gear. It matters little whether the items make sense or not. Thus, they are looking for correlations and not making causal inferences.

Part of the problem with this analysis, besides confusing correlation and causation, is that if we are looking at correlations between swimming gear and a million other products, then almost certainly some correlations will be statistically significant due to chance alone. When we assess significance, we usually set a threshold in order to infer a real correlation. Often, we set the threshold so that just one time in twenty we will infer a nonzero correlation when the true correlation really is zero. But if we are testing 1,000,000 correlations with this standard, about 5%, or one in twenty, will be significant just by chance. This is true even if all 1,000,000 correlations are truly zero. But 5% of 1,000,000 is 50,000. Even if we raise the threshold to 1 in 100, or even 1 in 1,000, there will still be several of these *spurious* correlations.

GFT did an excellent job of estimating the influenza rate for previous years. Predicting the past, though, is much easier than predicting the future. "It is hard to predict, especially the future" is a famous quote variously attributed to the physicist Neils Bohr and to the Hall of Fame baseball player/philosopher Yogi Berra. It is one thing to make data fit the past well; prediction of what happens next is a different matter.

Beginning in 2009, the GFT began to deviate somewhat from the CDC data and then it substantially underestimated the H1N1 pandemic. GFT estimated a rate of approximately 0.04, when the true rate was 0.08 at the peak. This outbreak of the flu occurred in the summer, not the winter like most outbreaks, which suggests that some of the search terms might be "winter" related, rather than "flu" related. For this summer outbreak, there seems to have been changes in search behavior in two of the categories: influenza complications, and terms for influenza.

From 2011 through 2013, GFT then overestimated the flu rate. Out of 108 consecutive weeks, the GFT model overestimated the flu rate 100 times, while Lazer et al. found that very simple models that use the CDC data (and only the CDC data) could do as well as GFT. Such models use some of the most recent values to predict the current value of the flu rate. Lazer et al. concluded that:

> *For example, by combining GFT and lagged CDC data, as well as dynamically recalibrating GFT, we can substantially improve on the performance of GFT or the CDC alone.... This is no substitute for ongoing evaluation and improvement, but, by incorporating this information, GFT could have largely healed itself and would have likely remained out of the headlines.*

Shortly after being launched, it was discovered that media coverage of influenza affected search behavior, which in turn affected the GFT estimates, which, as illustrated in Figure 6.1, affected the media coverage in a feedback loop.

Another problem, present from the start, is that internet searches may not be related to influenza at all. Most people who go to their doctor with flu-like symptoms do not have any type of flu. This number varies, but it is often around 90%.[7] These people believe (enough to go see their physician

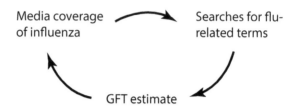

Figure 6.1 Feedback loop with searches, GFT, and media reports

and maybe get tested) that they have the flu, but they don't usually. If their behavior suggests that they believe they have the flu, then their internet searches are likely to involve flu terms that will be picked up by GFT.

Google discontinued GFT in 2015.

6.2 SOCIAL MEDIA

On the platform Twitter, there are over 500 million Tweets per day on various topics, including health, where there are over 100 million users who Tweet daily. The sheer magnitude of this information makes it amenable to monitoring for all sorts of things, including disease prevalence, where people will often reveal through Twitter many of their health issues, even embarrassing maladies such as diarrhea or sexually transmitted infections. Furthermore, Twitter data can be obtained in real time from an internet site and Twitter data can be purchased directly from Twitter or from a Twitter service provider. With this, as with Google searches, researchers can monitor the frequency of particular terms, such as "fever," "cough," or "flu."[8]

Dr. John Brownstein from Boston's Children's Hospital and Harvard Medical School is one of the leaders in the new field of *web epidemiology*. Brownstein has participated in a number of projects that used Twitter to monitor various aspects of public health. He is a co-founder of HealthMap,[9] a project that consolidates various data sources to infer the status of infectious diseases.

On January 13, 2010, a massive earthquake of magnitude 7.0 hit the country of Haiti. While it destroyed many structures and much of the infrastructure, internet and cell phone access was mostly intact. This allowed people to communicate to authorities the local conditions, both in terms of damage and health. Twitter and Facebook became the ways to alert authorities where problems remained. Using these platforms, together with the automated surveillance system HealthMap, epidemiologists were able to track the frequency of "cholera" mentions on Twitter.[10] They also found that "as the official number of cases increased and decreased, so did the volume of informal media reports about cholera."[11] In addition to tracking the disease spread, the social media platforms were also helpful "for connecting ground responders to each other."

In 2014, as an epidemic of Ebola ravaged west Africa, Tweets were being monitored for an increase in the frequency of terms related to the disease.[12] Beginning on July 24, 2014, Tweets began to warn of the presence of Ebola. For example, one Tweet read "Guys,#EbolaVirus is in Lagos. Be informed. Be careful." Over the next seven days the number of Tweets regarding Ebola increased dramatically. Tweets on July 24 reached approximately 1.2 million people and by July 30 the number had risen to over 120 million, a hundredfold increase. The warnings on July 24 were three days before an official news alert, and seven days before a CDC announcement. Lawal Bakare, a Nigerian dentist, created @EbolaAlert, a Twitter campaign to spread information about the disease.[13] Because those who have died from Ebola still carry the virus, it is possible to spread the disease through funerals. One of Dr. Bakare's Tweets involved "How to conduct safe & dignified burial of a patient who died from suspected or confirmed Ebola (EVD)." By garnering a substantial Twitter following, Dr. Bakare and his @EbolaAlert are credited with helping to contain the 2014 Ebola outbreak in west Africa.

Other health applications of Twitter data abound. Researchers at the Commonwealth Scientific and Industrial Research Organisation (CSIRO) in Australia have developed

WeFeel (`http://wefeel.csiro.au`), a system that infers the emotion (surprise, joy, love, sadness, anger, and fear) in a Tweet. They can use this to monitor the emotional health of a region.

Brownstein also studied side effects of 23 medications. He and his colleagues found 60,000 Tweets related to drug side effects. Many of these described the types of side effects that were encountered. Twitter has also been used to track outbreaks of foodborne disease. Often, it is possible to drill down to find the source of the contamination.

6.3 PREDICTION MARKETS

Public health officials, physicians, nurses, pharmacists, and others often have informed opinions about the state of a disease and its future over the course of a few months. Prediction markets have been designed to extract and quantify this "crowd sourced" information. We begin with the idea of futures markets and how they work. Prediction markets are similar but there is no tangible product associated with them.

Futures markets allow traders to agree to a transaction of some commodity at some future date at a specified price. For example, suppose oil is today selling at about $53/barrel. A headline at `cnbc.com` says "Oil Market is about to Get 'Ugly'" which suggests that oil prices may continue to drop because US production is high. If I believe this, and I would like to capitalize on it, I could enter a prediction market and try to purchase a futures contract on one barrel of oil. Another trader, who believes that oil prices will stay stable (or even increase in price), may offer to buy one barrel from me one year from today for $55. Of course, I don't own a barrel of oil (or a barrel of anything, for that matter), so one year from today I will have to buy one barrel of oil on the world market so I can sell it to the other trader for $55.

Since I believe oil prices will be lower in one year, I would be inclined to enter into this agreement. But, another trader may be willing to offer me $56 for that one barrel of oil. I would look over the market to find the highest bid for my one barrel of oil which I will be required to sell one year from today. If $56 is the highest bid I receive, then that bidder and I would enter into an agreement. If the price of oil rises to $60/barrel in one year, I will be obligated to purchase one barrel of oil on the world market for $60 so I can sell it to the bidder for $56. I lose $4 on the deal because I thought the price would go down, when it really went up. If the price of oil were to drop to $48/barrel in a year then I would simply buy the one barrel of oil on the market for $48 and my partner in the contract would be required to purchase it from me at the agreed upon price of $56. I would make a profit of $8.

Thus, if the price goes down, I make a profit, and if the price goes up, I incur a loss. This is just the opposite of "investing" in a commodity.

Futures markets such as this have led to prices of commodities that are more stable in markets such as agriculture and husbandry. Within the last few decades markets have arisen to predict future events that do not involve any tangible product. These are called *prediction markets*.

For example, the winner of the United States presidential election can be predicted from such a market. These markets operate as follows. I can buy one share of "Donald Trump wins the popular vote for president" (or "Hillary Clinton wins the popular vote for president"). If this event occurs, then the value of my share is $1.00; if the event does not occur, then the share is worth nothing. Suppose that on October 1, 2016, I believe that Trump has a small, but not tiny, chance of winning the popular vote. Maybe, I believe that Trump's chance of winning is 20% and Clinton's chance is 80%. This implies that no other candidate, in my opinion, has any chance at all of winning. I might find someone willing to sell one share of "Trump" for $0.10. If I really believe Trump has a 20% chance of winning then the expected value of my investment in one share of "Trump" is

$$E(\text{one share of Trump}) = \$0 \times 0.80 + \$1 \times 0.20 = \$0.20.$$

The expected value of one share of "Trump" is \$0.20 (under my assessment that Trump's probability of winning is 0.20) and someone is willing to sell me one share for \$0.10! If Trump were to win the popular vote (which he did not) then my \$0.10 investment would turn in to \$1.00. If the conditions were to change, making it less likely that Trump would win, then there would be few (or no) bidders willing to pay \$0.10 for a share and the price would go down.

Prediction markets, which require the investors to put their money "where their mouth is," work well for the following reasons:[14]

1. they effectively collect information from various participants which often have different sources of information,

2. the possibility of financial reward is an incentive to reveal knowledge,

3. they provide feedback to other participants through the changes in prices, and

4. they guarantee that information is shared anonymously.

Prediction markets are regulated by the Commodity Futures Trading Commission. Currently, the only prediction market in the US that allows participants to use their own money (under certain circumstances) is the Iowa Electronic Markets (IEM) established in 1988 by the Tippie College of Business at the University of Iowa.

The IEM has set up a prediction market regarding the spread of the influenza virus, commonly known as "the flu." Participants were recruited from the medical, pharmaceutical, and public health communities in Iowa. Of the 160 recruited only 64 participated. Applicants must be approved before being allowed to invest. Rather than investing their own money, participants were given "H1N1\$100" (i.e., *funny money*). The reward for correct prediction was therefore not financial; instead, high account balances reflected a certain mental acuity.

Table 6.2   Some of the questions in the Iowa Electronic Markets 2009 influenza market.

How many cases of novel influenza A (H1N1) will be confirmed in the United States by the end of May 2009?
200 or fewer cases
201 to 350 cases
351 to 600 cases
601 to 1100 cases
1101 or more cases

How many states in the United States will have at least one confirmed case of influenza A (H1N1) by the end of May 2009?
1 to 10 states
11 to 20 states
21 to 30 states
31 to 40 states
41 to 50 states

How long will the novel influenza A (H1N1) outbreak last?
Before May 31
June 1 to June 30
July 1 to July 31
After July 31

Some of the questions that were the basis of contracts are shown in Table 6.2. Although there are just three questions listed in the table, there were really fourteen types of contracts or investments that one could make. For example, if I believed that the spread of H1N1 would be low, I might invest in the "200 or fewer cases" or possibly in both the "200 or fewer cases" and the "201 to 350 cases." If the actual number of confirmed cases was, for example, 150, then the first of these investments would be worth $1.00 on May 31, 2009. If, as was actually the case, the number of confirmed cases exceeded 1,100, then both of these investments would be worthless. Notice how *specific* these scenarios are: "novel influenza A (H1N1)" not just "the flu," "confirmed cases" not

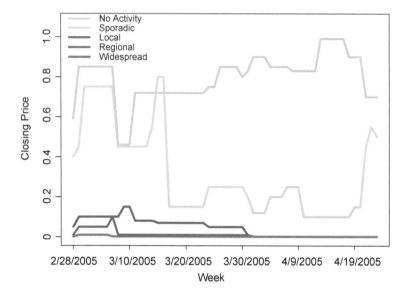

Figure 6.2 Price of shares of each "stock" for various levels of spread of influenza.

just "probable," "in the US" not "worldwide," and "by the end of May 2009" not just "winter/spring 2009." A vaguely worded question could create a dispute about whether the event occurred.

In previous years, the IEM had run prediction markets for other questions regarding influenza. Figure 6.2 shows the closing price of one share of each "stock" regarding the intensity of novel influenza A (H1N1) in week 16 (April 17 to April 23) of 2005. A color system was used to describe the spread of influenza. The system uses the following categories (taken directly from their web site):

Yellow No Activity: No laboratory-confirmed cases of influenza and no reported increase in the number of cases of ILI.

Green Sporadic: Small numbers of laboratory-confirmed influenza cases or a single influenza outbreak has been reported, but there is no increase in cases of ILI.

Purple Local: Outbreaks of influenza or increases in ILI cases and recent laboratory-confirmed influenza in a single region of the state.

Blue Regional: Outbreaks of influenza or increases in ILI and recent laboratory-confirmed influenza in at least two but less than half the regions of the state.

Red Widespread: Outbreaks of influenza or increases in ILI cases and recent laboratory-confirmed influenza in at least half the regions of the state.

For example, one share of "Green" is worth $1.00 if the actual level of influenza is green (sporadic) in that week; otherwise, the share is worth nothing. As time passed, the spread of influenza was apparently not as bad as originally thought, so shares of "red," "blue," and "purple" (respectively, "Widespread," "Regional," and "Local") quickly became worthless as investors moved away from these. The highest prices were for shares of "No Activity" or "Sporadic."

Notes

[1] See https://searchenginewatch.com/2016/08/08/what-are-the-top-10-most-popular-search-engines/.

[2] Ginsberg, J., Mohebbi, M.H., Patel, R.S., Brammer, L., Smolinski, M.S., and L. Brilliant (2009). Detecting Influenza Epidemics Using Search Engine Query Data, *Nature*, 457, 1012-4.

[3] *Ibid*, p. 1012.

[4] Lazer, D., Kennedy, R., King, G., and Vespignani, A. (2014). "The parable of Google Flu: traps in big data analysis." *Science*, 343(6176), 1203-1205.

[5] *Ibid*. Also see http://blogs.scientificamerican.com/observations/why-big-data-isnt-necessarily-better-data/.

[6]Fung, K. (2014). Google Flu Trends' Failure Shows Good Data > Big Data, *Harvard Business Review*. Accessed online at `http://hbr.org/2014/03/google-flu-trends-failure-shows-good-data-big-data`.

[7]`https://www.forbes.com/sites/stevensalzberg/2014/03/23/why-google-flu-is-a-failure/`.

[8]Special care must be taken to exclude the use of these terms in cases where no actual disease is implied. For example the term "Bieber fever" refers to an obsession with the Canadian pop star Justin Bieber. Just searching for "fever" will catch all instances of "Bieber fever," however.

[9]See `www.healthmap.org/en/`.

[10]Hirschfeld, D. (2012). Twitter data accurately tracked Haiti cholera outbreak. *Nature.*

[11]*Ibid.*

[12]See Odlum, M., and Yoon, S. (2015). What can we learn about the Ebola outbreak from tweets? *American Journal of Infection Control*, 43(6), 563-571.

[13]Carter, M. (2014). How Twitter may have helped Nigeria contain Ebola. BMJ: British Medical Journal (Online), 349.

[14]Adapted from Nelson, F. D., and Polgreen, P. M. (2010). "Using Prediction Markets to Forecast Infectious Diseases." In Kass-Hout, T. and Zhang, X. (Eds.) (2010). *Biosurveillance: Methods and Case Studies*, (pp. 154-169). Chapman and Hall/CRC.

II

Disease Investigations

When surveillance indicates a potential increase in the incidence of some disease, epidemiologists jump into action. Their job is to find out the source or cause of the increase and how to mitigate or prevent future cases of the disease. In a very real sense, epidemiologists are like detectives investigating a crime. They seek out evidence in an effort to uncover the truth about a disease. Sometimes, as in the case of the 2001 anthrax attack, there really is a crime and epidemiologists work together with law enforcement to determine not just the cause of the disease, but who perpetrated the attack. This is an example of *forensic epidemiology*, which uses methods from the field of epidemiology to study and possibly determine causality in criminal or civil cases.

We begin with a description of the steps that an epidemiologist might take in a disease investigation. Not all of the steps are performed in every investigation, and sometimes it is necessary to cycle through several steps. The next chapter should give an overview of how epidemiologists go about finding the cause for a disease and taking action to mitigate the chance or severity of future outbreaks.

In the subsequent chapters, we discuss several examples of disease investigations. We begin in Malaysia with a traditional epidemiological investigation that determined the cause of a disease eventually called the Nipah virus. From there, we jump to the former Soviet Union and a suspicious smallpox outbreak in Aralsk, which is now part of Kazakhstan. Then we head to California and a syphilis outbreak in San Francisco and from there we look into the investigations surrounding the anthrax attacks that followed the terrorist attacks in 2001. Next we look at surveillance for a *noncommunicable* disease, cancer, in

Los Alamos, New Mexico. The last two chapters discuss the hunt for the causes of yellow fever and microcephaly, where cleverly designed studies were required to gain insight.

The common theme among these is how scientists used data to guide them in their search for the cause of a disease and the source of the outbreak or attack.

CHAPTER 7

Steps in Investigating an Outbreak

A<small>N</small> epidemiologist is a scientist who studies the incidence and prevalence of diseases and their source or causes.[1] When a disease outbreak occurs, a team of epidemiologists and other medical specialists is often sent to the affected region to investigate, and (it is hoped) to find the cause of the disease. This team is like a group of detectives. They seek evidence of any sort that might point them in the direction of a cause. This usually involves collecting data of some sort, as well as interviews with those who are infected and perhaps the health care workers who treat them.

Before we look at some case studies in the chapters to follow, here we first look at how to actually conduct an outbreak investigation, where we begin by outlining the steps an epidemiologist would take. Then we look at how we determine whether a patient has the disease being investigated. While you might think that's easy, it can actually be quite challenging depending on the difficulty of diagnosis, where it's always possible to fail to diagnose someone with the disease of interest (a "false negative" diagnosis) or to say someone has the disease when they actually do not (a "false positive" diagnosis). It's important to understand and account for these kinds of errors both in trying to understand the effectiveness of a diagnostic test and when we try to estimate the incidence of disease.

7.1 STEPS IN OUTBREAK INVESTIGATION

Mark Dworkin, in *Outbreak Investigations Around the World: Case Studies in Infectious Disease Field*,[2] describes the following steps in an outbreak investigation.

1. **Verify that an Outbreak is Occurring.** A surveillance system may signal that more cases of a disease are occurring than would be expected by chance. One of the most common causes for a surveillance signal, whether it be in disease surveillance, quality monitoring, or some other sort type of surveillance, is simply the misrecording of data. For example, if 15 cases are input as 51 cases, the system may raise an outbreak signal when none is occurring. If the number of cases is truly the number reported, then additional steps and testing should be done to assure that there really are sick people and an investigation is warranted.

2. **Confirm the Diagnosis.** Many diseases produce very similar symptoms. For example, Ebola and influenza can, initially at least, be confused because both exhibit similar symptoms. For example, the CDC lists the first four symptoms of influenza as

   - fever or feeling feverish
   - headache
   - muscle or body aches
   - feeling very tired (fatigue)[3]

The CDC also lists the first four symptoms of Ebola as

- fever
- severe headache
- muscle pain
- feeling very tired (fatigue)[4]

The next several symptoms that the CDC lists are different, but you can see why in the early stages that Ebola and influenza can be confused. Both diseases are serious, but the case fatality rate for Ebola is much higher, often exceeding 50%. Often collected samples must be shipped off to a well-equipped laboratory (sometimes the CDC in Atlanta). For outbreaks of foodborne diseases, such as *Salmonella*, a lab can confirm the suspected agent and can even identify the strain of the bacterium.

3. **Assemble an Investigation Team.** For a small outbreak of a nonlethal disease, an individual can be dispatched to investigate. In more serious cases, however, a team of experts with varying expertise may be sent. This team may include one or more epidemiologists who have expertise in dealing with the disease being investigated. Others might include a statistician (who would design data collection strategies and analyze the resulting data), a person knowledgeable in medical records (if many medical records must be examined), an environmental scientist (if the problem might be related to environmental factors), a computer scientist (if victims used the computer and internet to arrange contact), and others depending on the nature of the investigation.

4. **Create a Case Definition.** A clinical case definition usually involves a combination of symptoms, their duration, and possibly exposure to other infected people or some other source. The clinical case definition does not involve laboratory testing. The laboratory criterion for

diagnosis usually involves identifying the agent or antibodies to the agent. Determining the criteria for the clinical case definition often involves issues of sensitivity and specificity (see Section 7.2), where there is an inherent trade-off. We want the clinical definition to be broad enough that it includes all or most of those with the disease, but not so broad that it mistakenly includes many who don't have it.

5. **Case Counts.** Once it is determined which people have the disease, it is necessary to assess the extent of the problem and to identify how many have the disease. Then the process of identifying the source of the problem can begin.

6. **Perform the Epidemiologic Analysis.** In most cases data, beyond the presence/absence of the disease, are collected. This will often include demographic data (gender, race, age, etc.), geographic data (where do the patients live, work, or go to school), exposure data (what activities did the patients participate in), etc. Analysis of this data may suggest how the infected individuals acquired the disease. Epidemiologists are like detectives who are chasing evidence. Analysis of data, together with personal interviews (of patients and health care workers), can lead the investigators to the problem's source.

7. **Perform Supplemental Laboratory or Environmental Investigation.** In some situations, laboratory or environmental analyses beyond that done on individual persons are required. For example, in the investigation of a foodborne illness, specimens from a restaurant may be collected and analyzed. For an airborne agent, analysis of ventilation ducts or cooling towers may be done.

8. **Develop Hypotheses.** Scientific knowledge is obtained by the iterative process of developing and testing

hypotheses. The data, including interviews, will often point to a particular cause and the investigators will make a hypothesis. As more information comes in, it should be evaluated in the context of this hypothesis. If the additional data are inconsistent with the hypothesis, then the hypothesis must be discarded or revised.

9. **Introduce Preliminary Control Measures.** Control measures might include things like closing a restaurant, recalling packaged foods, cleaning air ducts, isolating patients, prescribing prophylactic antibiotics, etc. These can be serious measures; for example closing a restaurant can have long-term implications for the owners and its employees. This must be weighed against the possible damage to public health if the restaurant remains open.

10. **Decide Whether Additional Studies are Required.** If there is still uncertainty about the source, additional studies may be required. Investigators must decide what additional information would be helpful in identifying the source.

11. **Perform Additional Studies to Determine the Cause.** In many situations a *case-control* study is warranted. For a case-control study several *cases* (i.e., those people who have the disease) are identified and compared to a control group. The control group is a set of persons who are similar in demographic characteristics but who don't have the disease. What is it that distinguishes those who have the disease from those who don't? If this question can be answered, then the investigators have a lead in the search for the cause or source of the disease.

12. **Perform New Control Measures and/or Ensure the Compliance of Existing Control Measures.** When the source is identified with reasonable certainty, control measures can be determined and put into place and followed.

13. **Communicate Prevention Information and Findings.** For a high profile disease investigation, the news media will be anxious for information and answers. The information conveyed to them must be accurate and honest, while at the same time preserving privacy. Names, and sometimes even exact locations, or patients cannot be divulged. Health care workers also deserve their privacy. To the extent possible the media and the public deserve to know the "who, what, when, where, why, and how" of the disease outbreak. Finally, the results of the disease investigation may be submitted to a scientific journal for publication. This is especially important for new or emerging diseases, such as Ebola, Zika, Nipah, etc.

14. **Monitor Surveillance Data.** This brings us back to square one. It is likely that the outbreak was initially detected by routine disease surveillance. Once the outbreak has been thoroughly investigated, the source identified, and control measures put into place, the problem reverts back to one of disease surveillance.

Dworkin points out that these steps need not necessarily be done in this order. Some steps may be skipped, while others are going on simultaneously. Some of these, particularly the hypothesizing and collecting additional information, may cycle.

7.2   DETERMINING WHO HAS THE DISEASE

To determine whether a patient has the disease being investigated, physicians rely on laboratory tests or on case definitions. Often both are available, in which case the lab test is usually more accurate. For pertussis, commonly known as whooping cough, both exist. The CDC and the Council of State and Territorial Epidemiologists have established the following definitions:[5]

Clinical Case Definition

A cough illness lasting at least 2 weeks with one of the following: paroxysms of coughing, inspiratory "whoop," or posttussive vomiting, without other apparent cause (as reported by a health professional).

Laboratory Criteria for Diagnosis

Isolation of *Bordetella pertussis* from clinical specimen.

Positive polymerase chain reaction (PCR) for *B. pertussis*.

Case Classification

Probable: meets the clinical case definition, is not laboratory confirmed, and is not epidemiologically linked to a laboratory confirmed case.

Confirmed: a case that is culture positive and in which an acute cough illness of any duration is present; or a case that meets the clinical case definition and is confirmed by positive PCR; or a case that meets the clinical case definition and is epidemiologically linked directly to a case confirmed by either culture or PCR.

Setting the bar for a clinical case at two weeks seems arbitrary. As is, it creates a definition that will identify most cases of pertussis, but it will wrongly identify many people as having pertussis when they don't (a *false positive*). Changing the criterion to one week will identify more pertussis patients, but more people who don't have the disease will test positive. Changing it to three weeks will do the opposite: fewer false positives, but less ability to detect actual cases (i.e., more *false negatives.*)

Often, a laboratory test involves measuring the amount of something in the blood. In cases like this, a cutoff must be determined to separate those who we infer *have* the disease, from those we infer *don't*. Of course, for some diseases, like anthrax, the presence of a single organism is enough to infer the presence of the disease.

This leads to the concepts of *sensitivity* and *specificity*. Sensitivity is the probability that a patient tests positive for the disease given that he or she has the disease. Mathematically,

$$\text{sensitivity} = \Pr(+ \text{ test} \mid \text{Disease}).$$

Here, the vertical bar "|" indicates conditioning, and is read "given." The specificity of a test turns this around; it is the probability that a patient tests negative given no disease; that is

$$\text{specificity} = \Pr(- \text{ test} \mid \text{No disease}).$$

Contrary what many people believe, there is no single number that could be considered the *accuracy* of a test. There is an accuracy for those who have the disease, called the sensitivity, and there is an accuracy for those who do not have the disease, called the specificity. We would like for both the sensitivity and specificity to be 1 (or nearly so) because a probability of 1 indicates a sure event. For nearly all tests, the sensitivity and specificity are less than 1, often quite a bit less. Changing the cutoff in the case definition or laboratory test can change the sensitivity, but only at the expense of the specificity, and vice versa. For example, we could change the case definition to include a cough for one week (from two weeks). This would increase the sensitivity, because those with pertussis would be more likely to test positive. On the other hand, it would decrease the specificity because many of those without pertussis would test positive.

If the probability distributions of the continuous measurements in the diseased and non-diseased groups do not overlap, then it is possible to determine a cutoff that discriminates the two groups with near certainty. Consider the PSA (prostate specific antigen) test given to men to test for prostate cancer. Higher levels of PSA are associated with prostate cancer. Prostate biopsies (examination of tissues snipped from the

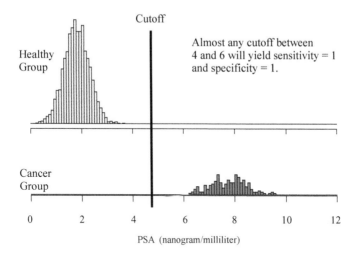

Figure 7.1   An ideal situation where those with the disease and those without have distributions that do not overlap. The sensitivity and specificity are both virtually 1.

prostate) are often given after high PSA levels are discovered through a blood test. Figure 7.1 illustrates the situation where the two groups do not overlap. Any cutoff between 4 and 6 will clearly separate the two groups. In this case the sensitivity (the probability of detecting the disease when it is present) and the specificity (the probability of saying no disease where the disease is not present) are both virtually 1.

In most practical problems, the two distributions overlap, making it difficult to infer the presence or absence of the disease. Figure 7.2 illustrates this situation. The trouble is that no single cutoff will guarantee that someone is in the diseased or non-diseased groups. If we wished to set the bar so that no one who is disease-free will be told they have the disease, then a cutoff of 6 would work, as shown in Figure 7.2. The trouble is that about half of those with prostate cancer will be missed. For these men, no treatment (possibly lifesaving) would be given. In this case the test has low sensitivity, but high specificity.

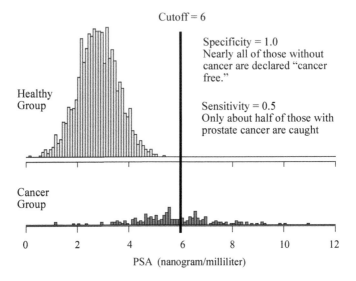

Figure 7.2 Using a cutoff of 6, most of those who do not have the disease are cleared, but many who do have the disease are missed. The specificity is high (nearly 1) but the sensitivity is low.

If we wanted to increase the chance of detecting the disease, we could move the cutoff to the left, and if we wanted to make the sensitivity nearly equal to 1 (meaning that we would catch all those with prostate cancer), then we would have to set the cutoff at about 1 as shown in Figure 7.3. This choice would infer that nearly all those with prostate cancer will be detected, but many men without prostate cancer will also be detected. For those disease-free men who are told that they have prostate cancer, there are costs that cannot be overlooked: the cost and discomfort of a biopsy, the anxiety of believing they might have cancer, etc.

At this point you could imagine dragging the cutoff line to the left or the right in these figures. Dragging it to the right will increase specificity but decrease sensitivity. In other words, as the cutoff moves to the right we will decrease the rate of false positives (a good thing) but increase the rate of

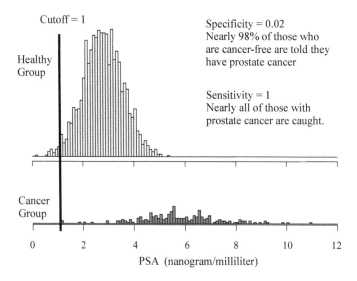

Figure 7.3 With a cutoff of 1, nearly all of those with the disease are caught, but many who do not have the disease are erroneously told they have the disease. The sensitivity is high (virtually 1) but the specificity is low.

false negatives (a bad thing). Conversely, dragging it to the left will increase sensitivity but decrease specificity. That is, moving it to the left will decrease the rate of false negatives (good) but increase the rate of false positives (bad). The choice of where to place the bar for a test will ultimately depend on the relative costs, not just in dollars, of false positives and false negatives. The common value used today is 4, which yields a sensitivity of 0.90 and a specificity of 0.93. See Figure 7.4. Is this a good compromise? Ask your physician.

7.3 DETERMINING THE PROBABILITY OF HAVING THE DISEASE

The discussion so far has been from the point of view of the disease investigators, not from the point of view of the patient being tested. A person being tested is interested in the

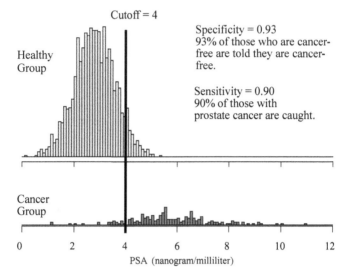

Figure 7.4 With a cutoff of 4, both the sensitivity and specificity are 0.90 or higher.

probability of having the disease conditioned on the test being positive; in other words

$$\Pr(\text{Disease} \mid + \text{ test}),$$

or the probability of not having the disease given a negative test

$$\Pr(\text{No Disease} \mid - \text{ test}).$$

These two probabilities are called the *positive predictive value* (PPV) and *negative predictive value* (NPV), respectively, and they represent the probability of making the correct diagnosis given the result of the test. While these may look like the probabilities for sensitivity and specificity the conditioning is reversed.

For example, the sensitivity is $\Pr(+ \text{ test} \mid \text{Disease })$, while the PPV is $\Pr(\text{Disease} \mid + \text{ test})$. Notice how the conditions are reversed. If I test positive for a disease, what I want to know is "How likely is it that I have the disease?" Analogously, if I test negative, I want to know "How likely is it that I am really disease free?"

We can determine these probabilities by reasoning through a large hypothetical population. To illustrate this, consider HIV testing where just one in ten thousand (0.01%) of a low risk population (i.e., not men who have sex with men or intravenous drug users) is truly infected. This is often called the *base rate*. The sensitivity for HIV testing is approximately 0.999 and the specificity is approximately 0.9999.[6] With such high numbers for sensitivity and specificity it would seem that the PPV and NPV would be large as well. In a population of, say, 100,000,000 people, 0.01% (a proportion of 0.0001) will have HIV; this amounts to 10,000 people. Also, 99.9%, or 99,990,000, will not have HIV. Among those who truly have HIV, the number who test positive will be

$$\text{sensitivity} \times \text{number tested} = 0.999 \times 10,000 = 9,990.$$

The remaining 10 will test negative. Among the 99,990,000 who do not have HIV, the number who test positive will be

$$(1 - \text{specificity}) \times \text{number tested}$$
$$= (1 - 0.9999) \times 99,990,000$$
$$= 9,999$$

The remaining 99,980,001 will test negative. These calculations, which are illustrated in Figure 7.5, indicate that out of the original 100,000,000 in the population, $9,990 + 9,999 = 19,989$, will test positive. (These are underlined in the figure.) Among these $19,989$ who test positive, only 9,990 actually have the disease. The PPV is thus

$$\text{PPV} = \frac{9,990}{19,899} = 0.4998.$$

This means that if you test positive for HIV, the probability is only about one half that you actually have the disease. Most people are amazed that the PPV is so low. Remember, this comes from a sensitivity of 0.999 and a specificity of

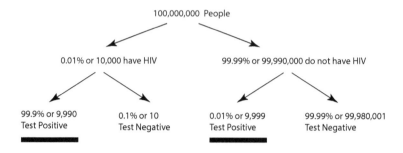

Figure 7.5 Calculations for a hypothetical population of 100 million people at low risk.

0.9999, both very high numbers. This paradoxical result can be explained by the sheer number of those who do not have HIV. Even a tiny fraction of the noninfected population of 99,990,000 will be a sizable number. Also, since the disease is rare, there are few in the population who have it, so very few test positive. Among those who test positive only about half actually have the disease.

The NPV can be computed as well. This is

$$\text{NPV} = \frac{99,980,001}{10 + 99,980,001} \approx 0.9999999.$$

Thus, you can be fairly certain that you don't have HIV if you test negative.[7]

7.4 DETERMINING THE CAUSE OF A DISEASE

Determining causation can be difficult in any field, but is especially challenging in epidemiology. This is because there is randomness in who gets a disease and who does not. Many who are exposed to a risk factor do not get the disease, and many who are not exposed still get the disease. This is related to the ideas of *necessary* and *sufficient* causes (see the "A Closer Look" box in this section).

A Closer Look

Necessary and Sufficient Causes. A cause is *necessary* if it must be present when the effect occurs. In other words, if the suspected cause is absent then the effect never occurs. A cause is *sufficient* if the effect must always occur whenever the cause is present. We can summarize these as they apply to the study of diseases with some simple if-then propositions:

Necessary – If the disease is present, then the cause must have occurred.

Sufficient – If the cause is present, then the disease must follow.

Aside from some genetic conditions, where the presence of some genetic abnormality inevitably leads to a disease, sufficient causes are rare in the study of disease. *Mycobacterium tuberculosis* must be present in persons with tuberculosis, but the bacterium can live dormant in some people who show no symptoms of the disease. Most diseases, however, have neither necessary nor sufficient causes because sometimes the causes occur in those with the disease and sometimes those with the disease have not experienced the purported cause.

For example, some people who smoke cigarettes get lung cancer, but nonsmokers can also get lung cancer. Thus, smoking is not a necessary cause for lung cancer. There are also people who smoke their whole lives but don't get lung cancer, so smoking is not a sufficient cause for lung cancer. For most diseases, exposure to some possible cause does not inevitably lead to the disease. We must look closer at the kind of data and evidence that could lead us to infer a cause and effect relationship.

An old maxim in statistics is that "correlation does not imply causation." For example, shoe size is correlated with reading ability in every elementary school. It's not that larger shoe sizes cause better reading (or vice versa); in this case there is another variable that affects both shoe size and reading ability. The variable is *age*; older children have larger feet

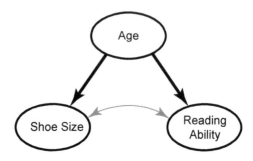

Figure 7.6 A causal diagram for the relationship between reading ability and shoe size.

and read better. A variable like this that affects both variables under consideration is called a *confounder*. A causal diagram is often used to display associations between variables and the possible causal relationships. Figure 7.6 illustrates this idea and shows the relationships among the variables. Thick lines indicate accepted causal relationships. Thin gray lines with double arrows indicate observed associations that are not causal. (Dashed lines indicate preliminary or questionable causal relationships, but there are no dashed lines in Figure 7.6.)

In general, there are several criteria that can support an inference of causation. These include the following.

1. **Association** – The first criterion is *association* or *correlation*. There must be an observed relationship between the possible cause and the disease. By itself, this doesn't establish causation but without some relationship the absence of any correlation or association makes it difficult to make a causal inference.

2. **Time Order** – The cause must come before the effect. Obviously, if disease D occurs before the variable C, we can't say that C causes D.

3. **Consistency** – If multiple studies suggest that the factor C is present whenever disease D occurs, we have

consistency. Again, by itself this doesn't prove causation, because there could be other variables (the confounders) that affect the likelihood of disease in *all* studies.

4. **Elimination of Other Possible Explanations** – When doubt is cast on all other possibilities, this supports the case of causation.*

5. **Coherence** – Is there a coherent scientific explanation of the causal explanation? If so, this bolsters the inference of causality. A dose-response relationship, whereby higher doses or exposures lead to a higher probability of the disease, would indicate a causative effect. If the exposure causes similar outcomes on *in vitro* experiments, then this suggests causation.

To illustrate these criteria, let's address the question of whether smoking causes lung cancer. Many studies have demonstrated an association between smoking and lung cancer, and smokers have a much greater chance of developing lung cancer than nonsmokers. Rises in lung cancer have occurred several decades *after* an increase in smoking; this is true for both men and women, but for women the increase in smoking and lung cancer both lagged that of men by twenty years. Per capita smoking was nearly zero in 1900, but rose steadily from 1900 until 1963, with a temporary drop during the Great Depression. Lung cancer rates for men began increasing in 1930 and peaked in 1990; for women, the increase occurred about 1960 and peaked in 2002. The relationship between smoking and lung cancer has been observed wherever it is studied. There is a relationship between the number of cigarettes smoked and the risk of lung cancer, and stopping smoking subsequently reduces the risk of lung cancer by about 40%.

---

*We are reminded of the quote by the character Sherlock Holmes in Arthur Conan Doyle's novel *The Sign of the Four:* "How often have I said to you that when you have eliminated the impossible, whatever remains, however improbable, must be the truth?"

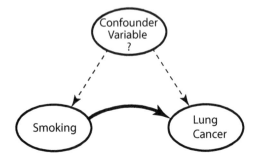

Figure 7.7 A causal diagram for possible confounding variables in the relationship between smoking and lung cancer.

It has been proposed that genetic factors can act as confounding variables; that is, they affect both the likelihood of smoking and of getting lung cancer. These theories, or any other theories of why smokers are at an increased risk of lung cancer, have never been proven, suggesting that smoking is the cause. Figure 7.7 shows a causal diagram for this possible relationship.

7.5 STUDY DESIGN

How a study is designed affects the ability to make causal inference. Here is a list of some of the common study designs used in epidemiology in increasing order of their ability to reach a causal inference.

1. **Cross-Sectional Study** – In a cross-sectional study, a population is identified and a sample from it is selected for the purposes of making inference about that population at that time. For example, epidemiologists may want to estimate the proportion of a population who have been vaccinated. Since a cross-sectional study is just a snapshot of a population in time, it cannot indicate causation.

2. **Case-Control Study** – We will see several examples of case-control studies in the rest of this book. Here,

researchers identify a set of *cases*, that is, people who have the disease. They then try to match the cases with *controls*, that is people who do not have the disease. Usually, matching is done on as many demographic or environmental variables as possible. Sometimes matching is made on a one-to-one basis, where one control is matched with one case. In some case-control studies there are several controls matched with each case. In other cases, the control group is selected in the aggregate to be as similar as possible to the case group. Researchers then probe into the possible exposures that subjects had in the time preceding the disease. Inference is made regarding the difference in exposure between cases and controls. For example, we might reach the conclusion that cases are significantly more likely to have been exposed to environmental condition E than controls. Researchers would like to conclude that those who have been exposed to E have a greater chance of getting disease D, but they must be content by concluding that those with disease D are more likely to have been exposed to E. Case-control studies can suffer the problem of *recall bias*, which occurs when those with the disease (cases) recall more possible exposures than those who do not have the disease (controls).

3. **Prospective Cohort Study** – In a prospective cohort study, researchers identify a cohort of people and then follow them forward in time. The possible exposures are noted and after a sometimes lengthy period of time (often years) the disease rates are compared. Estimates of the risk of getting a disease can be obtained and compared across various exposures.

4. **Randomized Control Clinical Trial** – This is the gold standard of clinical research. In such an experiment researchers are able to impose a treatment on subjects. The simplest trial often involves just two treatments, a new experimental procedure (drug, surgery,

intervention, etc.) and a placebo. (A placebo is a "nothing" treatment against which the result of the new procedure is compared.) Usually, the assignment is done at random and neither the subject nor the person administering the treatment knows which treatment the person is receiving. Such trials are said to be *double blind.* Because treatments are randomly imposed, any difference in the response (for example, the disease rates) must be due to the treatment. There are many situations where clinical trials are impossible because it is unethical to impose potentially harmful conditions on people who volunteered for the study. For example, we could not conduct an experiment where we imposed smoking on one group and nonsmoking on the other group.

In the coming chapters we will see some of these designs applied to disease outbreaks. We will also see how epidemiologists respond and how they seek to understand the cause of the disease so they can minimize the chance of it reoccurring.

Notes

[1] The prevalence of a disease is the proportion of those in a population who *have* the disease, whereas the incidence is the proportion who *contract* the disease in a fixed time period. For chronic diseases, such as diabetes, many people have the disease; this is the prevalence. In a given period of time fewer people are diagnosed with the disease.

[2] Dworkin, M.S. (2010). *Outbreak Investigations Around the World: Case Studies in Infectious Disease Field,* Jones and Bartlett Publishers, Sudbury, MA.

[3] See www.cdc.gov/vhf/ebola/pdf/is-it-flu-or-ebola.pdf.

[4] *Ibid.*

[5] Source: www.cdc.gov/mmwr/preview/mmwrhtml/rr5503a3.htm.

[6]See Gigerenzer, G. (2002). *Calculated Risks: How to Know When Numbers Deceive You*, Simon and Schuster, New York.

[7]Bayes' Theorem is the tool that statisticians usually apply to reverse the conditions in a conditional probability. What we have done here involves essentially the same calculations as Bayes' Theorem, but we have used proportions of a large group rather than individual probabilities.

CHAPTER 8

The Nipah Virus

OFTEN it's not sophisticated statistical methods but investigative "shoe leather" and basic statistics that are required to figure out the cause of an outbreak. This is a story of such an outbreak. In September 1998 in the peninsular part of Malaysia near the city of Ipoh, a mysterious disease began showing up in local residents. Symptoms included fever and headache, along with drowsiness and sometimes convulsions. Most of the victims were pig farmers or those who handled pigs in some way. Pigs in this part of Malaysia were getting sick from the disease in alarming numbers, although the fatality rate was lower than it would prove to be in humans.

8.1   INITIAL SUSPECT: JAPANESE ENCEPHALITIS

Initial testing found antibodies from Japanese encephalitis (JE), which is usually not lethal to humans. This new disease, however, was highly lethal in human victims. Also, finding antibodies to a disease is less definitive than finding a live virus in a person. Presence of an antibody suggests that the person has had the disease sometime in the past, and it was possible that many of those testing positive for JE had had it previously.

JE is a vector-borne disease, meaning that it is carried by some animal, usually an insect, from one person or animal to another. In the case of JE (and also yellow fever and Zika), the vector is the mosquito. Because of the findings of antibodies to JE in some of those who had this new disease, it was believed that mosquitoes were carrying the disease from pigs to humans. The solution was to apply mosquito fogging to kill as many of the mosquitoes as possible, in order to reduce the chance of people being bitten by infected mosquitoes. In this case fogging did little to lessen the spread.

There were a few holes in the argument that the new disease was JE. First, the fatality rate of the new disease was quite high, often around 50%; JE is not nearly so lethal. Second, most of the victims were pig farmers and ethnic Chinese, in a country where Chinese made up less than a quarter of the population. It would seem that if mosquitoes carried the disease, a broader spectrum of the population would be affected. Third, with just a few exceptions, most victims were adults. Mosquitoes, if they were truly the vectors, would not be so discriminatory as to bite adults only.

An outbreak that occurred slightly later in the village of Kampung Sungai Nipah, also in Malaysia (see Figure 8.1), exhibited much the same disease spread. However in this case samples from a man with the new disease were tested by Paul Chua from the University of Malaya Medical Center. The analysis of this sample, conducted partly in the United States, indicated that this was not JE, but rather a new viral infection. It was called Nipah virus encephalitis, after the village that contained the first confirmed case. The background rate, or noise, was virtually zero because this was a newly discovered disease. Thus, the occurrence of a single case is a signal of an outbreak. In the Malaysian outbreaks, 283 people were infected and 109 died, for a case fatality rate of about 39%.[1]

For many viruses, there is an animal that carries the disease even though the virus does not make the animal sick. Such an animal is called the *reservoir host* for the virus. For Nipah, the pigs were not the reservoir host because they were getting sick from the disease. The pigs must have been getting the disease from some other source, likely the reservoir host.

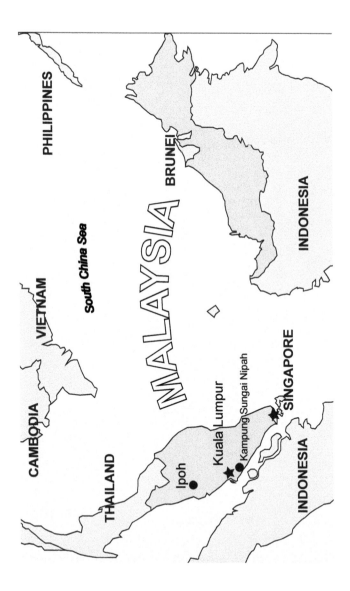

Figure 8.1 Map of Malaysia. The village of Kampung Sungai Nipah is on the western edge of peninsular Malaysia, southeast of the capital Kuala Lumpur.

We say that pigs were the *amplifier host* because they apparently contracted the disease from the reservoir host (whatever animal that might be), and were sickened by it before passing it on to humans. It still remained to discover what animal was the reservoir host.

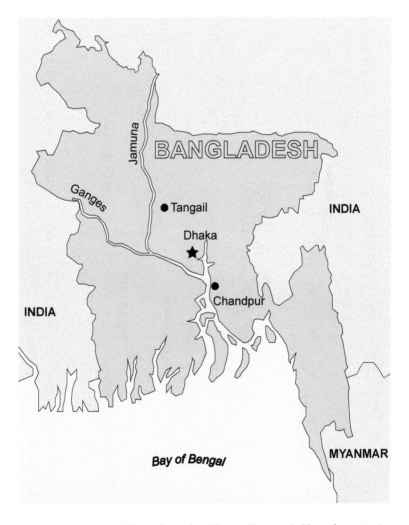

Figure 8.2 Map of Bangladesh. The village of Chandpur is in southern Bangladesh and Tangail is northwest of Dhaka, the capital.

8.2 THE NIPAH VIRUS RETURNS

In 2001, Nipah arose again, but this time in Bangladesh (see Figure 8.2), a predominantly Muslim country with very little pig farming. It began in Chandpur, a village of approximately 600 people in southern Bangladesh, where there were thirteen cases and nine fatalities. Unlike what happened in Malaysia, the disease seems to have spread through person-to-person contact.

The next outbreak occurred in January of 2003, one hundred miles north of Chandpur, in the Naogaon District. In January 2004, it appeared again, near Dhaka, the capital of Bangladesh. Interestingly, outbreaks in Bangladesh tended to occur in January and, unlike in Malaysia, most cases were children, particularly boys.

A team of investigators, including an American from the Centers for Disease Control and Prevention (CDC), arrived to determine how people were infected. The epidemiologists designed and conducted a *case-control* study.

Statistically Speaking

A *case-control study* compares people who have a disease (the cases) with people who do not have the disease (the controls). The study then compares the two groups looking for differences in an attempt to identify the source of the disease.

The statistical analysis of these types of case-control studies can range from the fairly simple to highly sophisticated. What researchers are looking for is clear demographic, behavioral, or medical differences between the cases and the controls. At its most basic, it could just be counting in order to identify a factor that is present in all or nearly all of the cases and absent in all or nearly all of the controls. At its most complicated, it could involve very sophisticated statistical methods to try to identify very subtle, but perhaps critical, differences between the cases and controls.

For this case-control study, epidemiologists developed a questionnaire about risky behaviors that might have led the victims to the reservoir host, or some other contact. They asked about a number of behaviors that might have put children at risk: handling dead animals, eating certain foods, fishing, hunting, contact with sick people, etc. It is not enough that one of these behaviors is associated with the disease. The question is really did they do it more often than those in the control group. Thus, even though those with the disease might have handled dead animals, it could (and was) the case that the controls handled dead animals to the same extent. The only behavior that was statistically significant was "climbing trees." The significance of this had to wait until the next outbreak.

Another outbreak occurred in January 2005 in the Tangail District, 60 miles from Dhaka. Again a case control study was performed and the most significant behavior found was drinking date palm sap: many of the cases had drunk date palm sap, and most of the controls had not. Date palm sap is a sweet thick juice obtained from date palm trees, much like maple syrup is obtained from maple trees.

8.3 SOLVING THE MYSTERY

Further investigation into how date palm sap is collected yielded some clues. High on date palm trees, the bark is stripped away and a "straw" usually made of bamboo is drilled into the tree as a tap. A clay pot is then placed below this tap, and overnight the sap runs into the pot. See Figure 8.3. The resulting product is sold early the next day as a sweet drink. This process occurs usually in December and January, the coldest months of the year in Bangladesh, and is consistent with the timing of the outbreaks. Since the top of the pot is open, it can collect more than just the intended date palm sap. Often the sap contains insects, and bat urine or bat feces. In some cases, the pots were found to contain dead bats! This suggested that bats might be the reservoir host. Further

Figure 8.3 Date palm sap is extracted from the tree by drilling a "straw" into the tree and placing a small clay pot beneath it to catch the sap. Bat waste that was deposited into the sap in the pot contained the Nipah virus which subsequently infected humans who drank the date palm sap. © Shutterstock Images.

testing of the bats, including the testing of bat urine, indicated that bats do carry the Nipah virus, but are not sickened by it, confirming that bats are indeed the reservoir host for Nipah.

The bat as reservoir host is consistent with the observations in Malaysia, where the disease seems to have come from pigs. Many of the pig farms in Malaysia also had fruit trees near the pig sties. Fruit eating bats will come at night to eat the fruit, dropping pulp, along with urine or feces. The pigs eat some of the pulp remains or drink some of the water that was contaminated with bat droppings. In Malaysia, the spread seems to have been from bats to pigs to humans. By contrast, in Bangladesh, it was directly from bats to people.

This knowledge was put to use to prevent future outbreaks. In Malaysia, many fruit trees were removed from the areas near pig farms. In Bangladesh, screens were constructed and placed around the clay pots to keep the bats out. These simple measures have likely prevented additional outbreaks, but some outbreaks have continued to occur. According to the World Health Organization, from 2001 to the present there have been 280 cases in Southeast Asia and 211 fatalities, for a case fatality rate of over 75%.[2]

Unfortunately, there is as yet no vaccine for either humans or animals and the primary treatment for human cases is intensive supportive care.[3]

Notes

[1] Quammen, D. (2012). *Spillover: Animal Infections and the Next Human Pandemic.* WW Norton & Company.

[2] *Ibid.*

[3] For additional information, see `www.cdc.gov/vhf/nipah` and `www.oie.int/fileadmin/Home/eng/Media_Center/docs/pdf/Disease_cards/NIPAH-EN.pdf`.

CHAPTER 9

Smallpox and the Aralsk Incident

SMALLPOX is a viral disease caused by the *variola major* virus. First symptoms, which begin ten to fourteen days after exposure, include fever, headache, and body aches. This is followed by a rash (see Figure 9.1) that often begins on the face or head, and spreads throughout the body; this is the most contagious stage. The sores from the rash become pustules which eventually scab and fall off. The fatality rate for this form of smallpox is about 30% but survivors are left with disfiguring scars. Waves of smallpox have devastated many parts of the world since at least as far back as the sixth century.

9.1 SMALLPOX VACCINE

It was widely noticed in sixteenth century rural England, especially on dairy farms, that having contracted cowpox, usually from dairy cows, renders a person immune from smallpox. Cowpox is related to smallpox, but is much milder.

In the spring of 1776, Edward Jenner (1749–1823) acted on this suspicion. He scraped material from a sore on the hand of a young woman infected with cowpox, and injected it under the skin of an 8-year-old boy. Several weeks later, he

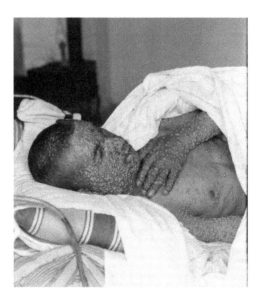

Figure 9.1   Pockmarked face and body of a man with smallpox.
© Shutterstock Images.

deliberately exposed the boy to smallpox. The boy did not acquire smallpox; the exposure to cowpox seemed to make the boy immune. This is one of the earliest recorded clinical trials. An experiment like this, where someone was deliberately exposed to a potentially deadly disease, could never be run today.

Jenner's discovery led to the first vaccine, and it formed the foundation for the science of immunology. His experiment was not the first to theorize that exposure to cowpox would render a person immune from smallpox. Two decades earlier, a farmer (not a physician) named Benjamin Jesty performed the procedure of injecting cowpox material into uninfected people. Jesty's experiment differed from Jenner's in that Jesty did not subsequently expose the subjects to smallpox. As a fellow Englishman, and one of the founders of the field of statistics, Francis Galton (1822–1911) said, "In science credit goes to the man who convinces the world, not the man to whom the idea first occurs."[1]

The *Variola* virus that causes smallpox is one of the few viruses that is not a *zoonosis*.[2] The fight to eradicate a disease from the earth is difficult because most infectious diseases can retreat into the animal population, only to emerge later. For example, the Nipah virus, discussed in Chapter 8, resides in bats. Even when a Nipah epidemic is stopped the virus still exists in the bat population, which can infect humans at some later time. Since smallpox occurs only in humans (as far as we know), once it is eradicated in the human population, it would theoretically be gone for good. The last naturally occurring case of smallpox was a Somali man who contracted smallpox (and recovered!) in 1977.[3] In 1980 the World Health Assembly declared smallpox eradicated, the first, and to date the only, disease to be eliminated. Smallpox had been one of the greatest killers in recorded human history. As late as 1967 there were approximately 10 to 15 million cases of smallpox with a fatality rate of about 25%.[4]

After 1977 efforts were made to move all remaining *Variola* virus into two highly secure BSL-4 labs.[5] One of these is at the CDC in Atlanta, Georgia, and the other is at the Research Center for Virology and Biotechnology in Koltsove, Russia. Since smallpox had spread virtually worldwide, there is no guarantee that all samples of the virus have been collected. It is possible that additional samples exist in a hospital's or doctor's freezer somewhere in the world. In fact, in 2014 six vials of freeze-dried *Variola* virus were discovered in a cardboard box at a National Institutes of Health facility in Bethesda, Maryland.[6]

Today, the smallpox threat still exists, but (almost certainly) only through the possibility of a bioterrorist attack. In other words, the stable rate is zero and the noise is zero, so a single case should produce a signal. If the virus could be aerosolized, that is, made to float in air for some period of time, and if it could be delivered to a heavily populated area, then a widespread epidemic could occur. The nearly two-week incubation period, together with the high mobility of people in modern cities, means that the epidemic could spread across the country or the world in short order.

9.2 SMALLPOX OUTBREAK AT ARALSK

In the late summer and early autumn of 1971, several people were diagnosed with smallpox in the port city of Aralsk at the northeast corner of the Aral Sea in what was then the south central part of the Soviet Union. The Aral Sea sat between the republics of Kazakhstan in the north and Uzbekistan in the south. These have been independent countries since the breakup of the Soviet Union in late 1991.

Today, the Aral Sea looks much different than it did in 1971. If you looked at a map from today, you would see a much smaller lake than the one in 1971. The Soviet Union began using the water from the two major rivers that flow into the Aral Sea in irrigation projects, and as a result, the flow into the lake was reduced significantly. Figure 9.2 shows the outline of the lake in the early 1970s compared to today, with Lake Michigan shown alongside. The Aral Sea was once the fourth largest lake in the world, larger than Lake Michigan. Since 1960 the surface area has been reduced from 68,000 km^2 to 14,280 km^2.[7]

On September 22, 1971 two people were diagnosed with smallpox in Aralsk. One was a 23-year-old woman who fell ill on September 10 with symptoms that included "general discomfort, weakness, and headache."[8] Within a few days she had fever, dizziness, headache, nausea, and vomiting. After she developed a bright red rash on September 20, she was diagnosed with measles. Physicians began to suspect smallpox when vesicles formed on some of the papular eruptions that formed on her extremities. She died on September 27 of a form of smallpox called hemorrhagic smallpox. With this form, there is bleeding under the skin as it detaches from the underlying surface. The fatality rate from hemorrhagic smallpox is virtually 100%. She had not been vaccinated.

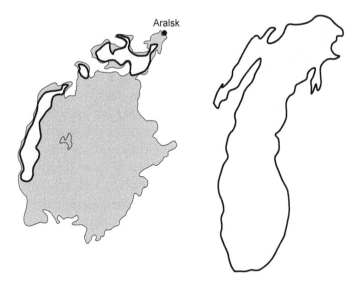

Figure 9.2 What remains of the Aral Sea today. Blue region indicates the remnant pieces of the Aral Sea. The gray region indicates the Aral Sea of the early 1970s. A map of Lake Michigan drawn roughly to scale is shown on the right.

Another patient at the same hospital, a 36-year-old woman began feeling ill, also on September 10 and with similar symptoms. She went to an outpatient clinic on September 13 where she was diagnosed with pneumonia and given antibiotics. She was admitted to the hospital on September 22 and was seen by the same physician as the previously admitted woman. Both women were diagnosed with smallpox on the same day by the same physician. Both women were previously diagnosed with other diseases (measles and pneumonia). Many diseases have similar symptoms at the onset of illness, making it difficult to make the correct diagnosis. Since the Soviet Union had not seen a case since 1961, smallpox was not high on the list of possibilities in 1971. The second patient had been vaccinated for smallpox, and survived.

Once people were diagnosed with smallpox, the local epidemiologists tried to find the source of the outbreak, and they enacted policies to restrict or eliminate the spread of the disease. One approach was to study all of the contacts that the smallpox patients had with previously ill persons. Do (or did) any of these have smallpox? If yes, from where did *these* patients contract smallpox? For a serious disease like smallpox, an extensive and thorough search is called for, and officials did a house-to-house canvass looking for cases. In an attempt to protect those in the community, officials encouraged vaccinations; such a strategy of vaccinating all of those around cases is now called *ring containment vaccination.* Officials can also use isolation and quarantine: patients with a confirmed case are isolated from all others, and those who have been exposed (i.e., who *might* have the disease) have restrictions placed on their movement. More about the Soviet officials' response at the end of this chapter.

The next patient to be diagnosed with smallpox was a 5-year-old boy whose case was discovered by the door-to-door canvass. The boy had been vaccinated and survived. According to the official Soviet report, which was written in Russian and not released until 2001 and not translated into English until 2002,[9] ten patients were diagnosed with smallpox. Of the ten, three died.[10]

The search for previous cases led back to a 9-year-old boy who had been visited on August 31 by his teacher. School began on September 1, and with the boy's illness, the teacher came to his house, probably to talk about missed work during the first week of school. The boy was admitted to the hospital on September 28, having already been ill for 32 days. He was in satisfactory condition and did not complain about any symptoms. He had become ill on August 27, earlier than any of the previously diagnosed cases. He felt "generally bad" with a headache and fever.[11] He developed a rash on August 30 and was seen by a doctor who diagnosed urticaria, or hives. He was prescribed tetracycline and aspirin, which are useless against smallpox. He recovered within a few weeks and returned to school on September 13.

But from whom did he contract smallpox? It turns out that his older sister was a 21-year-old marine biologist, the youngest researcher on the ship *Lev Berg* as it made a research expedition to collect data throughout the Aral Sea, which by 1971 had been shrinking for ten years, but it was still, at the time, a large lake. The ship left Aralsk on July 15 and returned on August 11, after making stops at Uyaly (sometimes spelled Uyalu), Komsomolsk-on-Ugyurt, and Muynak. Figure 9.3 shows the path of the *Lev Berg* in July/August 1971. The 21-year-old marine biologist had been ill with many

Figure 9.3 The Aral Sea in the 1970s between the republics of Kazakhstan and Uzbekistan, which were then part of the Soviet Union. The smallpox outbreak occurred in the northeastern port city of Aralsk. The approximate path of the *Lev Berg* in July/August 1971 is shown.

of the symptoms of smallpox. It is here that the official report and her own accounts from a 2002 interview differ. According to the official report, she began feeling ill on August 6, en route back to Aralsk. She suffered from a fever, headache, and muscle aches. Once she reached her home in Aralsk, she developed a rash. She was seen by a doctor who prescribed aspirin and antibiotics.

But from whom did she contract smallpox? The official report contends that she must have acquired smallpox while visiting one of the port cities, probably Ulaly on or about July 29. In a 2002 interview with Dr. Alan Zelicoff of Sandia National Laboratories she said, "I didn't get off the boat in Uyaly, because anybody from the time will tell you, it was forbidden for Russian women to get off the boat in a port that was a non-Russian port. This was because the Kazaks hated the Russians and their women. It was simply too dangerous."[12] She also claims never to have disembarked at any of the ports. No other member of the crew fell ill, so it is unlikely that anyone who disembarked at Uyaly brought smallpox back to the ship. Also, there were no reported cases of smallpox in Uyaly, or in any of the ports. Smallpox is a very "noticeable" disease. If someone in any of the port cities had had smallpox, it would have almost surely been reported.

On July 29 or 30 the *Lev Berg* passed just south of Vozrozhdiniye Island in the west central part of the Aral Sea. The island, once known as Nicholas the First Island, was allegedly the site of a Soviet biological weapons laboratory. Housing for the workers was on the northeast side of the island; the laboratory was southwest of that, and testing was done at the very southern tip of the island. Winds in this part of the Aral Sea were almost invariably from north to south, making this remote island an ideal place to test biological weapons. It is suspected that open air tests of biological weapons were carried out on this island. Is it possible that such a test of a smallpox weapon could have been conducted which infected the 21-year-old biologist even though she was several kilometers (probably about 15 kilometers, or 9 miles) away? If this occurred, why was she the only one infected? She claimed in the 2001 interview that being the youngest researcher meant she would have to do the work of pulling up nets at night. She said that "everyone else below was drunk or asleep."[13] With a thirteen-day incubation period, with little variability, this meant that her symptoms would begin about August 11.

The evidence that the very first patient in Aralsk to contract smallpox picked it up from an open air test of a biological weapon is, so far, circumstantial. But then came a confession! Dr. Pyotr Burgasov, a Soviet scientist was quoted in November of 2001 as saying:

> On Vozrozhdeniye Island in the Aral Sea, the strongest formulations of smallpox were tested. Suddenly I was informed that there were mysterious cases of disease in Aralsk. A research ship of the Aral fleet came 15 kilometers away from the island (it was forbidden to come any closer than 40 kilometers). The laboratory technician of this ship took samples of plankton twice a day from the top deck. The smallpox formulation — 400 grams of which was exploded on the island — "got her" and she became infected. After returning home to Aralsk, she infected several people, including children. All of them died. I suspected the reason for this and called the Chief of the General Staff of the Ministry of Defense and requested that the train from Alma-Ata to Moscow be forbidden to stop at Aralsk. As a result, an epidemic around the country was prevented. I called [Yuri] Andropov, who at that time was chief of the KGB, and informed him of the exclusive recipe of smallpox in use on Vozrozhdeniye Island. He ordered that not another word be said about it. This is a real biological weapon! [14]

The government report claims that ten people in Aralsk contracted smallpox, and three of them died. In their book

Table 9.1 Vaccination versus type of smallpox for Aralsk outbreak of 1971.

	Nonhemorrhagic	Hemorrhagic	Total
Vaccinated	7	0	7
Not Vaccinated	0	3	3
Total	7	3	10

Table 9.2 Vaccination versus survival for Aralsk smallpox outbreak of 1971.

	Survived	Died	Total
Vaccinated	7	0	7
Not Vaccinated	0	3	3
Total	7	3	10

Microbe: Are We Ready for the Next Plague?, Zelicoff and Bellomo suggest that there could be many more than the report suggested. Seven of the ten victims had been vaccinated against smallpox, and these were the seven who survived. There were three cases of hemorrhagic smallpox, and these were the three that died. The data are summarized in Tables 9.1 and 9.2. Statistical tests provide evidence that a vaccination tends to protect against hemorrhagic smallpox (but not against smallpox itself).[15] Since all seven of those who were vaccinated and had nonhemorrhagic smallpox survived, and since smallpox usually has a fatality of about 30%, it seems as if the vaccination protects against death.

9.3 QUESTIONS AND MORE QUESTIONS

This story raises a number of serious questions about the Soviet, now the Russian, biological weapons program. It appears that the Soviets had weaponized smallpox by aerosolizing it, meaning that it can be suspended in air for some period of time. It can travel great distances; the *Lev Berg* was at least

15 kilometers from Vozrozhdeniye Island. Also, very small amounts are required to cause infection when the first patient was infected. Dr. Pyotr Burgasov in the quote given above says that just 400 grams, under one pound, were released. The data also suggest, but don't confirm, that the strain of smallpox is particularly virulent. Those who were vaccinated still contracted many of the classic symptoms of smallpox, but all of those vaccinated survived. All three of those who were unvaccinated contracted the hemorrhagic form of small-pox, which is nearly always fatal, and indeed was fatal for the three people in this group: the teacher, a 4-month-old girl and a 9-month-old girl (vaccinations were not done until age one).

The Soviet response to the virus was quick. On September 23, 1971, the day after the first two cases were diagnosed, a team of epidemiologists was formed with the task of determining "exactly how smallpox had been introduced into Aralsk."[16] Even before the team was established, the first two diagnosed patients were put in isolation, and those who came into contact with these patients were vaccinated and put in quarantine. Beginning the same day, unvaccinated people were not allowed to leave the city, although travelers were allowed to get vaccinated at the airport and train station. Trains passing through Aralsk were not allowed to stop. According to the official report, ten people had contracted smallpox in this outbreak, although some believe that the actual count was considerably higher. The Soviets claimed to have vaccinated over 42,000 people, almost 33,000 of them for the first time.[17] No other cases of smallpox have ever been reported in the region.

Notes

[1] Riedel, S. (2005, January). "Edward Jenner and the history of small-pox and vaccination." In *Baylor University Medical Center Proceedings*, Vol. 18, No. 1, pp. 21-25. Taylor & Francis.

[2] A zoonosis is a disease that can jump between humans and animals. A jump from animals to humans is often called a *spillover*.

[3] https://www.cdc.gov/smallpox/history/history.html.

[4] https://web.archive.org/web/20070921235036/http://www.who.int/mediacentre/factsheets/smallpox/en/.

[5] Biological laboratories are graded on the amount of containment. The lowest is level 1 and the highest level 4. Level 4 labs, called BLS-4 labs, are the most secure and guarantee the greatest possible containment.

[6] https://www.nature.com/news/forgotten-nih-smallpox-virus-languishes-on-death-row-1.16235.

[7] See Gaybullaev, B., Chen, S.-C.,and D. Gaybullaev (2012). Changes in Water Volume of the Aral Sea After 1960, *Applied Water Science*, **2**, 285-291.

[8] Tucker, J. B., and Zilinskas, R. A. (2002). *The 1971 smallpox epidemic in Aralsk, Kazakhstan, and the Soviet biological warfare program.* Center for Nonproliferation Studies, p. 35.

[9] The Russian report was released to the Center for Nonproliferation Studies of the Monterey Institute of International Studies in 2001, and only then was it translated into English and made widely available. See Tucker, J. B., and Zilinskas, R. A. (2002). The 1971 smallpox epidemic in Aralsk, Kazakhstan, and the Soviet biological warfare program. Center for Nonproliferation studies.

[10] Zelicoff, A. P. (2003). "An epidemiological analysis of the 1971 smallpox outbreak in Aralsk, Kazakhstan. *Critical Reviews in Microbiology*, 29(2), 97-108.

[11] Tucker, J. B., and Zilinskas, R. A. (2002). *The 1971 Smallpox Epidemic in Aralsk, Kazakhstan, and the Soviet Biological Warfare Program.* Center for Nonproliferation Studies, p. 37.

[12] Zelicoff, A. P., and Bellomo, M. (2005). *Microbe: Are We Ready for the Next Plague?*. AMACOM Div American Mgmt Assn, p. 106.

[13] *Ibid.* p. 108.

[14] *Ibid.* pp. 97-108.

[15] The appropriate statistical method for testing whether the variables are independent is the Fisher's exact test. This test of independence is rejected with a P-value of 0.0083 for both tables. The chi-squared test of independence yields a p-value of 0.0016, but the expected frequencies are too small for this to be reliable. We will discuss the chi-squared test in more detail in Chapter 10.

[16] Zelicoff, A. P. (2003). An epidemiological analysis of the 1971 smallpox outbreak in Aralsk, Kazakhstan. *Critical Reviews in Microbiology*, 29(2), 97-108.

[17] Tucker, J. B., and Zilinskas, R. A. (2002). *The 1971 Smallpox Epidemic in Aralsk, Kazakhstan, and the Soviet Biological Warfare Program.* Center for Nonproliferation Studies, p. 54.

Syphilis and the Internet

SYPHILIS, if untreated, can be a serious and deadly disease. Being a bacterial infection, it is, however, treatable using antibiotics, often penicillin. It is mostly spread through sexual contact. A significant challenge when treating diseases like syphilis is that the social stigma attached to sexually transmitted diseases can make it challenging to identify those who are at risk and those who may have been exposed.

The first symptoms of syphilis appear on average 21 days after infection, where the most common symptom is a round, painless sore. From six weeks to six months after exposure individuals develop a flat rosy-colored, non-itchy rash that usually covers the body including the palms of the hands and soles of the feet. At this point, other manifestations may include hair loss, genital warts, and flu-like symptoms. Latent stage syphilis begins when these symptoms disappear and the infection becomes dormant for up to 20 years. During the latent stage, the infection is still detectable by blood testing and it can be treated and cured, but any damage done to internal organs is irreversible. If the infection progresses through the latent stage without treatment, it enters a final stage whose side effects include blindness, loss of motor skills, dementia, and damage to the central nervous system and internal organs that typically results in death.[1]

10.1 AN OUTBREAK IN SAN FRANCISCO

After years of decline, cases of syphilis began to rise in San Francisco in 1999, particularly among gay men and men who have sex with men, where from 1999 to 2003 syphilis increased from 44 to 522 cases a year, and from 1999 through September 2004 a total of 1,730 cases of early syphilis were reported in San Francisco.[2] Figure 10.1 shows that the problem was worse for men, particularly gay and bisexual men.

All that was known in June and July of 1999 was that the San Francisco Department of Public Health (SFDPH) had received two new cases of syphilis in gay men where, during an interview process with them, the public health department learned that both had met many of their recent sexual partners in an internet chat room.[3] It thus became necessary to track down the sexual partners of these two men.

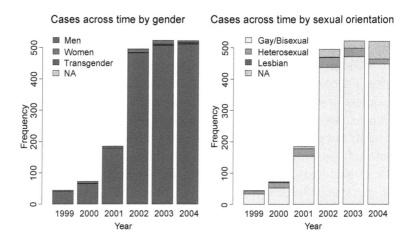

Figure 10.1 Epidemic curve of early syphilis by gender and sexual orientation, San Francisco, 1999-2004. Values for 2004 are only available through the first three quarters; the values shown are projections for the year.[4]

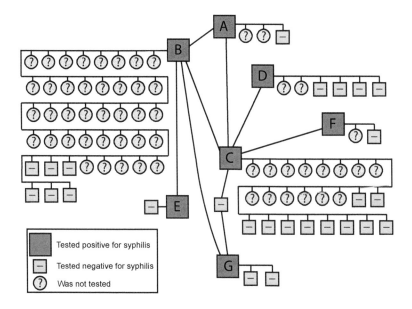

Figure 10.2 Actual exposure diagram. The boxes represent people and edges represent sexual contact.

This type of detective work often involves what is called an *exposure diagram*. Figure 10.2 is the actual exposure diagram, which graphically shows how people are linked, in this case via sexual contact. The red boxes indicate people who tested positive for syphilis, the green boxes indicate those who tested negative, and gray circles indicate those who have not been tested. An edge between vertices (boxes or circles) indicates known sexual contact. Here, seven persons tested positive and 26 tested negative; another 56 were not tested, so it is unknown whether they have syphilis. The exposure diagram gives the epidemiologists clues about how the disease has spread.

In the course of their investigation, the medical personnel had suspicions that many of the subjects had met new sexual partners online through an internet site. So, they conducted an analysis to better understand whether there was sufficient evidence to conclude that those who tested positive for syphilis were more likely to use the internet site to meet new partners. If the investigators' suspicions were correct, and men were meeting partners at a particular site, then it might be possible to warn those who had previously visited the site that syphilis rates were climbing so that they could get tested and receive treatment if necessarily. Also, it might be possible to post notices on the site about the syphilis outbreak and prevent future cases.

To do this, they conducted a case-control study. As we discussed in Chapter 8, the idea of case-control studies is to compare subjects who have a condition or outcome of interest (the cases) with a set of control subjects who do not have the outcome. Here the outcome of interest is testing positive for syphilis and the cases are six gay men with syphilis who reported to SFDPH in July and August of 1999 and the controls are 32 gay men without syphilis who presented to a city clinic in April through July of 1999.

The data are summarized in the 2-by-2 table in Table 10.1. Such a table shows the count for all four combinations of "Used Internet" (Yes/No) and "Contracted Syphilis" (Yes/No). Here we see in the column to the right of the table the six men who tested positive for syphilis and the other 32 tested negative, for a total of 38 subjects. In the row below the table, we see that ten of the 38 had used the internet chat room and 28 had not. And, within the table we see that of the ten who had

Table 10.1 Results of case-control study to determine whether those who test positive for syphilis have different internet habits than those who test negative. These are the *observed* counts in the data.

		Used Internet		
		Yes	No	
Syphilis	Yes	4	2	6
	No	6	26	32
		10	28	38

used the internet chat room, four tested positive for syphilis and six tested negatively. On the other hand, of the 28 who had not used the internet chat room, only two tested positive while 26 tested negative.

Table 10.1 is an example of a *contingency table*, and we often want to test whether the two variables in the table are related. In this case, we want to know whether participating in the internet chat room is associated with testing positive for syphilis. Intuitively, we have some reason to believe there is a relationship since in this data 40% (four out of ten) of those who used the internet chat room tested positive for syphilis, while only 7.1% (two out of 28) tested positive of those who did not use the internet chat room. But how to check this formally to see if the difference is large enough to confirm our suspicion? To do this, we do what statisticians call a *chi-squared* (pronounced "kai-squared" and is denoted as χ^2) hypothesis test.

Statistically Speaking

Table 10.1 is an example of a contingency table, that is, a table where each observation falls in one cell in a two-dimensional table. This table is 2×2 and covers all four combinations of "used internet" and "contracted syphilis."

We often want to know whether the proportions of those with a certain property are the same across two (or more populations). In this case, we are looking at whether the proportions who used the internet are the same for the syphilis and non-syphilis. This is because of the way we collected data. Recall that this was a case-control study where men with syphilis and without syphilis were selected. The random component was whether they used the internet. A test of whether the proportions are the same is done using a chi-square test of homogeneity. (The mechanics of this test are similar to that for a chi-square test of independence, although the assumption about which margin totals are fixed is different.)

To perform this test, we compute the expected frequency for each cell under the assumption of homogeneity. Without going into the details of the calculation, the table of expected counts, assuming the rate of internet use is the same for those who have syphilis and those who do not, is shown in the table below:

		Used Internet		
		Yes	No	
Syphilis	Yes	1.6	4.4	6
	No	8.4	23.6	32
		10	28	38

Under some conditions we will skip over here,[5] the χ^2 statistic is then the sum of the squared deviation between observed and expected divided by expected:

$$\chi^2 = \sum_{i=1}^{4} \frac{(O_i - E_i)^2}{E_i}.$$

In this case the χ^2 statistic is 5.98. If the proportions are really homogeneous, then the probability of getting a test statistic this large or larger is approximately 3 chances out of 100 (or 0.031). Since this is fairly unlikely, we conclude that the probability of internet use is not the same for the two groups. That is, we conclude that this is a difference in the syphilis infection rate between those who used the internet chat room and those who did not.

The details of this test, outlined in the Statistically Speaking example on page 137, are beyond the scope of this text. Doing the calculations, we find that assuming the true underlying internet use rates are the same for the syphilis and the non-syphilis groups, there is only a very small chance that we would observe the counts in Table 10.1 or something even more extreme. This probability works out to about 3 chances in 100, and so we conclude that there is sufficient evidence that the internet use rates differ.

10.2   THE INTERVENTION

Armed with this information, the epidemiologists and medical personnel attempted to identify the sexual partners of those who tested positive and to encourage them to get tested. This led to a clash between "the privacy rights of individuals with the need to protect public health"[6] where the internet service provider and the maintainer of the web site were unwilling to release the actual names or user names ("handles") of those who visited the site.

As the public health officials wrote,

> *A recent syphilis outbreak in San Francisco, Calif, challenged existing models of partner notification and community education. The outbreak occurred among gay men who met their sexual partners through an internet chat room. Because the partners had met in cyberspace, partner information was usually limited to handles (screen names). Moreover, the strongly held right to privacy of information accessible through the internet precluded us from directly learning the identity of partners through the internet service provider (ISP).[7]*

Since they were unable to contact the chat room participants directly, they took proactive steps such as posting notices, or ads, on the web site describing the recent outbreak. Specifically, an LGBT organization agreed to have a cadre of staff at their local organization post warnings about the syphilis outbreak and how it could be treated if detected. These postings were just "user" postings and were not advertisements. This staff monitored the web site for two full weeks and posted enough warnings that many users of the site became aware of the problem. Even though the investigators were not able to learn the identities of those who were most at risk, this solution did lead to a gradual reduction of syphilis in the community. It was likely that many who were at risk continued using the web site and thereby learned of their risk and sought appropriate medical help.

In addition, to increase syphilis awareness, education, and testing, the medical team funded a social marketing firm to develop a "Healthy Penis campaign." They wrote,

> *Based on that community input and expert opinion, the campaign focused on the secondary*

prevention of syphilis. Humor was selected as a key tool for conveying messages. The campaign included small media such as palm cards, T-shirts, comic books, and squeeze toys in the shape of Healthy Penis and Phil the Sore; posters in social venues and advertisements in community newspapers; and large advertisements on sides of buses, bus shelters, and in train stations; as well as news coverage in local and national print and broadcast media.[8]

For an example of one of their advertisements, see Figure 10.3.

Figure 10.3 An example of a "Healthy Penis" advertisement.[9]

Six months after the campaign began in June 2003, evaluations of the campaign showed that 80% of gay/bisexual men surveyed throughout San Francisco were aware of the epidemic. Furthermore, those who were aware of the Healthy Penis campaign were much more likely to know the symptoms of syphilis, how it is transmitted, and to have been recently tested for syphilis.

Notes

[1] See www.mayoclinic.org/diseases-conditions/syphilis/symptoms-causes/syc-20351756.

[2] Klausner, J.D., Kent, C.K., Wong, W., McCright, J., and M.H. Katz (2005). The Public Health Response to Epidemic Syphilis, San Francisco, 1999-2004, *Sexually Transmitted Diseases*, **32**, S11-18.

[3] Klausner, J.D., et al. (2000). Tracing a Syphilis Outbreak Through Cyberspace, *Journal of the American Medical Association*, **284**, 447-449.

[4] Source: *Ibid*, p. S12.

[5] Those familiar with chi-squared tests will note that this example does not meet the assumption that the expected counts are greater than five. Thus, to calculate the probability of getting a test statistic as large or larger than the one observed (known as the *p-value*), we used Monte Carlo simulation. Another appropriate test to use in this situation is Fisher's exact test.

[6] *Ibid*, p. 449.

[7] *Ibid*, p. 447.

[8] Klausner, et al., 2005, S14.

[9] Source: Ahrens K., Kent C.K., et al. (2006). Healthy Penis: San Francisco's Social Marketing Campaign to Increase Syphilis Testing among Gay and Bisexual Men. *PLoS Med* 3(12): e474. https://doi.org/ 10.1371/journal.pmed.0030474.

CHAPTER 11

The 2001 Anthrax Attack

THE disease called "anthrax" is caused by the spores from the bacterium *Bacillus anthracis*. Spores of anthrax occur naturally, but rarely infect humans. They are hardy and can survive in soil for up to 200 years. The incubation period, that is, the time from infection to first symptoms, is quite variable, being one day to two months. It is not typically communicable between humans. The most common exposure to anthrax occurs in people who work with animal hides or fur.

When anthrax spores enter the human body, they release a toxin (called the anthrax toxin) that causes the disease we call *anthrax*. Anthrax spores can be grown in laboratories and can be *aerosolized*, that is, made to suspend in air, making the powder a weapon. It is possible to manufacture fine powders that contain spores of anthrax, which enter the body and release the toxin. There are four types of anthrax infection for the different ways that the anthrax spores enter the human body. *Cutaneous*, or skin, anthrax occurs when the spores enter the body through small cuts or cracks in the skin. These cause a blister or lesion on the skin, but can spread. It is estimated that cutaneous anthrax is fatal in up to 20% of untreated cases, though with proper treatment most patients with cutaneous anthrax survive.[1] The second is *inhalational* anthrax, whereby the victim breathes in anthrax spores which settle in the lungs or lymph nodes and release the deadly toxin. This can take three to fourteen days. Symptoms are initially flu-like, but progress to shortness of breath, shock, and heart failure. With early aggressive treatment, the fatality rate is about 45%; without it the fatality rate is 85% to 90%.[2] Eating the meat of infected animals can cause *gastrointestinal* anthrax, which has a fatality rate of roughly 40% in properly treated patients.[3] Finally, anthrax can be injected directly into the blood stream, causing *injection* anthrax. According to the CDC, this is the rarest form of anthrax; no case of it has ever been reported in the United States, although there have been cases involving intravenous drug users in Europe. An outbreak in Scotland in 2010 involved 119 cases, 14 of whom died. Because it is so rare, estimates of its fatality rate are based on small samples.[4]

In 2001, shortly after the hijackings of four planes on September 11, anthrax spores in a fine powder were sent through the mail in at least five letters. This chapter chronicles the events as they unfolded that autumn. At first, there were many uncertainties about the cause of anthrax, and later about the type and source of the anthrax.

In terms of the signal and noise concept that we have discussed throughout this book, the noise for anthrax is virtually zero. There are almost no naturally occurring cases of anthrax, so a single case is a signal. The statistical concepts of sensitivity and specificity come into play as the diagnosis of anthrax is often missed. This chapter is mostly about the forensic epidemiology that was done to identify the source of the anthrax contamination and to clean up the affected buildings to make them safe.

11.1   THE FIRST SIGNS

On Sunday, September 30, 2001, Robert Stevens, a 63-year-old photo editor of the *Sun* tabloid published by American Media

Inc. (AMI), began having flu-like symptoms.[5] After being admitted to a hospital, he was diagnosed at first with meningitis. Only later was he diagnosed with inhalational anthrax. The next day, Ernesto Blanco, who worked in the mail room at AMI, was admitted to a hospital with similar symptoms. Once Stevens was diagnosed with anthrax, the CDC dispatched a team of epidemiologists to south Florida. Stevens died of inhalational anthrax on October 5, 2001; he was the first case of anthrax in the United States in 25 years. Where and how Stevens became infected was unclear. He was an avid outdoorsman and he had recently returned from a visit to North Carolina where he went fishing. It had been speculated that anthrax spores attached themselves to his clothes or he drank water from a contaminated spring. Acquiring anthrax in this way is quite rare. The same day that Stevens died, Blanco was diagnosed with anthrax.

Officials tried to allay the public's fears. Health and Human Services Secretary Tommy Thompson said, "I want to make sure that everybody understands that anthrax is not communicable, which means it is not spread from person to person. If it is caught early enough, it can be prevented and treated with antibiotics."[6] Later, President George W. Bush assured the public that this is "a very isolated incident."[7]

By October 5, the CDC team began swabbing various surfaces at AMI, looking for traces of anthrax. They found anthrax contamination on Stevens' keyboard and in the mailroom. The building was sealed off on October 7.

Although no anthrax-laced letter was ever found at AMI, it was suspected that a bioterrorist attack had occurred using the United States Postal Service (USPS) as a delivery system. By October 8, officials of the USPS and the unions that represented postal workers began to worry that postal workers might be exposed to anthrax while handling the mail. At the time, it was believed that anthrax could only be acquired by being exposed directly to the anthrax spores and that handling sealed letters was not a risk. This would later prove to be wrong.

On October 12, an assistant to Tom Brokaw at *NBC News* was diagnosed with cutaneous anthrax. She had noticed earlier (September 25) a pea-sized bump near her collarbone. On October 1, a physician at *NBC* suspected anthrax, but a test was negative. The woman followed the anthrax news carefully and believed her case could indeed be cutaneous anthrax. After all, she had recalled opening a powder-filled letter at *NBC*. It was not until October 12 that an immunohistochemical stain[8] on a biopsy was positive for anthrax. Shortly thereafter three more cutaneous anthrax cases were discovered at major news outlets. One was the 7-month-old son of a producer at *ABC* who had attended an office birthday party with his mother on September 28, although anthrax was not diagnosed until October 14. The second was an assistant to Dan Rather at *CBS News*. She routinely opened 2,000 to 4,000 letters per day and was diagnosed with skin anthrax on October 18. The third was an editorial assistant at the *New York Post*. When the anthrax-laced letter was discovered at *NBC*, there was, unfortunately, only a tiny amount of material left, so it yielded little information.

11.2   SOME EVIDENCE

Investigators' first bit of hard evidence came later, when on October 19 an unopened letter was found at the *New York Post* to contain powdered anthrax. There was a sufficient amount of the powder to yield information about the quality of the anthrax, and about its particular strain. Many believed that a person had to physically touch the material to get infected, but this letter was unopened. A New York City official said previously "... we see no public health concern."[9] This opinion was changing as it became clear that anthrax was being used as a weapon.

News reporters were not the only targets. On October 15, an intern at the office of Senator Tom Daschle opened a heavily taped letter and white powder was exposed. See Figure 11.1. The area in the Hart Office Building, which contained Senator Daschle's office, was closed. Antibiotics given before a

Figure 11.1 Envelope of the letter sent to Sen. Tom Daschle that contained high-grade anthrax. (Source: https:// archives.fbi.gov/archives/about-us/history/famous-cases/anthrax-amerithrax/the-envelopes)

person is symptomatic are effective; once symptoms begin, antibiotics are usually ineffective. Hundreds of employees at the Hart Office Building were given nasal swabs to detect anthrax spores, which were found in 28 people spread across the building. Testing positive on a nasal swab does not mean a person has been infected with anthrax. As a precaution, the antibiotic *Ciproflaxin* was prescribed to all those who requested it.

The powder from Senator Daschle's office was sent to the United States Army Medical Research Institute for Infectious Diseases (USAMRIID), a level 4 biosecurity lab (BSL-4) for analysis. In an article in the *New York Times* on October 24, Senator Daschle said, "it was a very strong form of anthrax, a very potent form of anthrax, which clearly was produced by someone who knew what he or she was doing."[10]

By mid-October, over 10,000 people were taking prophylactic antibiotics.[11] The intended recipients of letters were in the news media and on Capitol Hill, but these weren't the only victims. The anthrax-laced letters had to travel through the US postal system. Initially, it was believed that direct contact with the powder was required, but then postal employees were getting infected. On Tuesday, October 23, a postal worker from Hamilton Township, New Jersey, just east of Trenton where the anthrax letters were mailed, was hospitalized with a suspected case of inhalational anthrax. The next day six postal workers from DC were hospitalized with suspected anthrax; two of them would die from it.

By this time, the list of agencies participating in the disease investigation or the criminal investigation was long: FBI, USPS, Office of Homeland Security (not yet the Department of Homeland Security), CDC, EPA, OSHA, the Postal Inspection Service, and various local and state health and law enforcement agencies. These agencies were simultaneously conducting both a disease investigation and a criminal investigation. The disease outbreak involved finding and cleaning up places where the anthrax spores had been deposited. Anthrax is not generally spread from person to person, so setting up isolation or quarantine was not an issue. Because mail distribution centers seemed to be contaminated with anthrax spores, the problem of dealing with mail that was in the system and mail coming into the system had to be addressed in addition to the health problems. Eventually, several truckloads of mail were sent to Lima, OH, to be irradiated.

Post offices with anthrax contamination were found in New York, New Jersey, Florida, Washington, DC, and later Connecticut. The anthrax letters that had been found were mailed in New Jersey and had a Trenton, New Jersey, postmark. When a contaminated post office or mail distribution center was found, the set of facilities that traded mail with the contaminated station was determined. These facilities were then tested, and for those who tested positive, their set of trading facilities was determined. This process continued until all facilities tested negative. This process can be represented by a tree as shown in Figure 11.2. Nodes are USPS facilities (post offices, mail distribution centers, etc.) and edges indicate facilities that routinely traded mail. Red nodes indicate contaminated sites, and green indicates clean, or noncontaminated, sites. The tree continued branching until all ending nodes (called leaves in graph theory terminology) were green. The actual trees were much larger than shown in Figure 11.2;

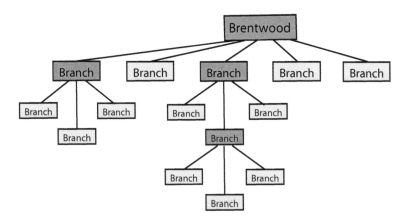

Figure 11.2 Hypothetical tree diagram for inspection of "trading partners" among post offices. Every trading partner of a post office that tested positive was also tested. Testing continued until all offices tested negative.

for example the Brentwood station in DC traded with over 50 stations, and the one in Trenton traded with over 150.

The contamination at so many postal facilities was at first perplexing. Health officials believed that direct contact with the white powder was necessary to become infected. As late as mid-October 2001, it was believed that sealed letters posed no risk. USPS senior vice president for government relations and public policy Deborah Willhite said "They [referring, presumably, to the CDC] said there was virtually no risk of any anthrax contamination in the facility, that without the letter being opened at Brentwood, there was no risk of any anthrax escaping, so neither the facility nor the employees needed to be tested."[12] By Sunday October 21, two Brentwood employees had contracted inhalational anthrax. One would die that day, and the other would eventually survive. A third Brentwood employee would die from inhalational anthrax the next day.

Since none of the letters were opened at the Brentwood facility, it became clear that anthrax spores could escape from a sealed envelope. Evidence also suggested that anthrax spores could hitch a ride on other pieces of mail that came in contact with the anthrax letters or the machinery that handled them. For the last two anthrax victims, this was the only plausible explanation. On October 31, a Bronx woman who worked at the Manhattan Eye, Ear, and Throat Hospital died from inhalational anthrax. She had no direct contact with people from the news media or with the mail distribution system. Three weeks later a 94-year-old Connecticut woman, who rarely left her home, also died from inhalational anthrax. Cross contamination of the letters was the most likely explanation.

By Monday November 12, investigators began combing through the congressional mail that had been accumulating. On Friday November 16 a heavily taped letter addressed to Sen. Patrick Leahy was uncovered. The letter was postmarked October 9 from Trenton, NJ, and it bore the same fictitious address with the childlike handwriting as the letter to Sen. Daschle. This still-sealed letter was a gold mine for those who were investigating the source of the anthrax and of the anthrax letters. It was estimated that the letter contained about one trillion anthrax spores per gram, enough to infect 100 million people. Such high grade weaponized anthrax was then believed to have come from a domestic source.

11.3   WINDING DOWN

By early December, the anthrax outbreak was winding down. There had been no new cases since the death of the Connecticut woman on November 21, and no new anthrax letters since one addressed to Sen. Patrick Leahy on November 16. Altogether, 18 anthrax infections had occurred. Eleven were inhalational anthrax and seven were skin anthrax. All five deaths (the AMI employee in Florida, the two Brentwood USPS employees in DC, the Bronx woman, and the elderly Connecticut woman) were from inhalational, not skin, anthrax.

Nearly 300 postal sites had been sampled and tested for anthrax contamination. The Postal Inspection Service had received nearly 20,000 reports of potential anthrax. Nearly all of these were false alarms, and many were hoaxes, as citizens began seeing white powder everywhere, in what was termed the "white powder hysteria."

11.4 WANING OF THE CRISIS

By Christmas the disease outbreak had waned, but the criminal investigation continued. The high-grade nature of the anthrax suggested a domestic source. Steven Hatfill, a scientist at USAMRIID at Fort Detrick, MD, became a person of interest. In the summer of 2002 FBI agents searched and swabbed his home in Frederick, MD. Hatfill vehemently denied any involvement in the anthrax infections, and in 2004 he sued the *New York Times* and the Justice Department. Although the suit against the *New York Times* was eventually dismissed, he received a \$5.8 million judgment against the government.[13]

Later, the investigation focused on another USAMRIID scientist, Bruce Ivins. The case against Ivins was largely circumstantial, based on data that showed a spike in overtime hours just before the anthrax attacks, and on the claim that he submitted false samples of anthrax to mislead investigators. Ivins died on July 29 in an apparent suicide.

Notes

[1] See www.cdc.gov/anthrax/basics/types/cutaneous.html.

[2] See www.cdc.gov/anthrax/basics/types/inhalation.html.

[3] See www.cdc.gov/anthrax/basics/types/gastrointestinal.html.

[4] An outbreak of anthrax among drug users in Scotland, December 2009 to December 2010. Glasgow: Health Protection Scotland (http://www.documents.hps.scot.nhs.uk/giz/anthrax-outbreak/anthrax-outbreak-report-2011-12. pdf).

[5] AMI, located in Boca Raton, FL, is also the publisher of the *National Enquirer*.

[6] See www.cnn.com/TRANSCRIPTS/0110/04/se.23.html.

[7] *The Washington Post* text of President Bush and German Chancellor Gerhard Schroeder news conference. Accessible online at www.washingtonpost.com/wp-srv/nation/specials/attacked/transcripts/bush_schroeder_text100901.html?noredirect=on.

[8] Immunohistochemical staining is a way to visually see when antibodies are activated, usually by linking them to an enzyme or a fluorescent dye which can then be seen under a microscope. The method has been widely used in medical analysis and diagnosis.

[9] http://www.cnn.com/2001/HEALTH/conditions/10/18/anthrax.CBS/.

[10] Source: www.nytimes.com/2001/10/17/us/nation-challenged-widening-inquiry-anthrax-mailed-senate-found-be-potent-form.html.

[11] Prophylactic antibiotics are intended to *prevent* a disease rather than to cure the disease.

[12] Howitt, A.M., Loenard, H.B., and D. Giles, eds. (2009). *Managing Crises: Responses to Large-Scale Emergencies*, CQ Press, p. 344.

[13] This involved $2.8 million in cash plus $150,000 annually for 20 years, which yielded a present value of about $4.6 million.

CHAPTER 12

Cancer in Los Alamos

So far, our examples have dealt with communicable diseases, but if there is a source that is causing increased incidence of some other disease, such as cancer, then this is an effect worth discovering. Knowing what environmental factors affect cancer rates helps us to create policies to mitigate the risk of cancer in the future. In this chapter, we look at the problem of trying to determine whether there is an unusual incidence of a chronic, non-communicable disease, namely cancer. As we will see, it's not easy to determine whether there is an environmental (or other) cause that potentially contributes to a greater cancer incidence. In fact, it's even difficult to figure out whether there is an increased incidence since the term cancer really covers a large family of diseases, all related to abnormal cell growth, and where there are more than 100 types of cancer.[1]

12.1   INITIAL SUSPICIONS

Los Alamos County is the smallest (by area) county in New Mexico and it is the home to Los Alamos National Laboratory, owned by the United States Department of Energy. The lab was created in 1943 for the purpose of designing and building an atomic bomb. Today the scientists at the lab do fundamental research in a number of scientific fields[2] including quantum information, high-energy physics, and bioinformatics.

In 1990 Tyler Mercier, a sculptor from Los Alamos County (LAC), worked as a volunteer on a project to monitor radiation around the town of Los Alamos. While he was performing this task he heard from several people who told him about clusters of cancer, especially brain cancer. His list of people in town who developed cancer grew after a public hearing in June 1991. Information collected in such a haphazard fashion is called *anecdotal evidence*. Such evidence is less reliable because it is subject to an observation bias. It is easy to notice high clusters of disease, but the vast majority of non-disease regions remain unnoticed.

Another issue that arises when observing clusters in both time and space is the *multiple comparisons* problem. When we do any single comparison of a disease rate in some region at some time against some standard, there is a small chance that we will discover an elevated rate simply by chance, even when none exists. However, when we test multiple regions, sometimes hundreds or thousands, for multiple time periods, there is a greater chance that we will discover at least one elevated risk. Add to this the fact that very often we are looking at many different types of cancer. For many of the communicable diseases studied so far a single case leads to a signal. For cancer, though, the background is noisy, making it difficult to detect a signal.

Statistically Speaking

If you pulled a coin out or your pocket and tossed it ten times you would be amazed if you got ten heads or ten tails. These events are very unlikely given that the coin is fair; the probability is $\frac{1}{1024}$ of getting ten heads, so the probability of getting all heads or all tails is twice this, or $\frac{1}{512}$, still very unlikely. In fact, if you got ten heads or ten tails, you would have doubts about the coin being fair.

There are 3,142 counties or county equivalents (Louisiana has parishes, not counties, and in Missouri, the city of St. Louis is not in any county) in the United States. Imagine if in each county one person was selected to toss a coin ten times. Since the probability is $\frac{1}{512}$ of getting all heads or all tails, we could expect that about this fraction of the 3,142 counties would see all heads or all tails. But $\frac{1}{512} \times 3142 \approx 6$ so we would expect about six counties to see this unlikely event. Each county that saw all heads or tails, and saw just this event, not events in other counties, would doubt that the coin is fair.

This is the issue faced by epidemiologists when looking at disease incidence across geographic regions. For example, cancer rates will vary across regions, and some will be at the very high end (like getting ten heads) and some will be at the very low end (like getting ten tails). If you add on to this the fact that epidemiologists will look at many different cancers over many different time periods, you will see that there is a very good chance of observing a high rate for some cancer, at some time, in some location. Whether such an observation indicates a truly elevated risk, or is just chance, is difficult to determine.

12.2 THE DATA

A report written in 1993[3] found mixed results in cancer incidence. Many of the cancers involved a very small number of cases in LAC. In such a small county, the difference between three and four cases can be significant. The authors of this study conclude, "Due to the small number of cases, random fluctuation in the county incidence could not be ruled out as causing the observed elevated rates."

Overall, the cancer incidence rate from 1970 to 1990 was about 75% higher than that observed in New Mexico and the US as a whole. Some cancers had higher rates: melanoma of the skin, thyroid cancer, prostate cancer, non-Hodgkin's lymphoma, ovarian cancer, and breast cancer. Others had rates that were lower than the reference populations (New Mexico and the United States as a whole): leukemia, lung cancer, and cancer of the digestive system. The most striking increase in

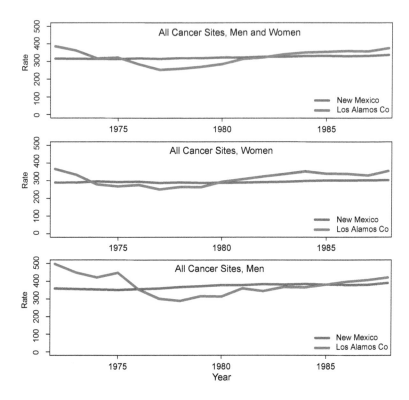

Figure 12.1 Five-year moving average for white non-Hispanics in New Mexico and LAC and for cancers of all sites: overall (top), women (middle), and men (bottom).

cancer incidence occurred for thyroid cancer that occurred in the mid- to late 1980s.

Figure 12.1 shows the age-adjusted five-year moving average cancer rates for white, non-Hispanic whites, including both men and women. Shown are the rates for New Mexico as a whole, and for LAC. Because New Mexico is so much larger, there is much less noise, so the rates are smoother. Being smaller, the rates in LAC jump around quite a bit. The figure indicates that overall cancer rates for LAC were higher in the early 1970s, then were lower until the early 1980s when they increased.

Statistically Speaking

Age-adjustment is used to make fair and realistic comparisons between regions. Suppose, for illustration, that one county (County A) is a popular retirement destination, where 30% of the population is retired, and therefore older. Suppose that another county (County B) is agricultural and contains many younger workers where just 10% of the population is retired. If the cancer risk was the same for both counties, we would expect that the raw cancer incidence rate would be higher for County B simply because its residents are older. If we looked at the cancer rates in each age category, say under 30, 30-50, 51-70, and over 70, we would expect higher rates for the older age brackets. County B would have a higher incidence rate because more of its residents fall in the upper age brackets. Age-adjustment involves using a standardized (hypothetical) population, say 30% under 30, 30% 30-50, 20% 51-70, and 20% over 70. The age-adjusted county rate would be

$$
\begin{aligned}
\text{Overall Rate} = {} & (\text{Rate in} < 30 \text{ group}) \times (\text{Prop in} < 30 \text{ group}) \\
& + (\text{Rate in 30-50 group}) \times (\text{Prop in 30-50 group}) \\
& + (\text{Rate in 51-70 group}) \times (\text{Prop in 51-70 group}) \\
& + (\text{Rate in} > 70 \text{ group}) \times (\text{Prop in} > 70 \text{ group})
\end{aligned}
$$

The age-adjusted rate is the rate that a region would have if each region had the same age make-up.

Variables other than age can also be considered for adjustment. Variables which could possibly affect the disease rate and vary across regions (such as counties) are called *covariates*.

Often data are reported as moving averages, such as a five-year moving average as in the LAC cancer data. This means that each value is the average of the rates from a given year along with the two previous years and the two subsequent years. For example, the plotted value from 1972 is the average of the values of 1970, 1971, 1972, 1973, and 1974. The next value, for 1973, is the average of 1971, 1972, 1973, 1974, and 1975. There is considerable overlap in consecutive years. As a result, the moving average produces smoother time series plots than plotting individual yearly values.

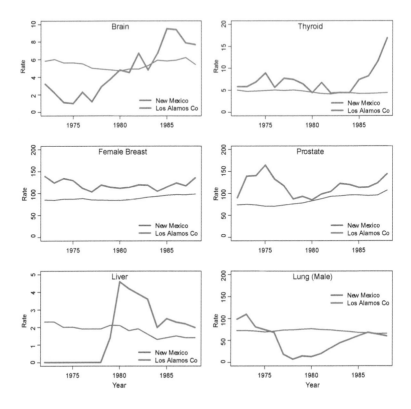

Figure 12.2 Five-year moving average for white non-Hispanics in New Mexico and LAC and for six specific cancers: brain, thyroid, female breast, prostate, liver, and male lung.

Figure 12.2 compares the cancer rates between New Mexico and LAC for a number of specific cancers. This figure only shows results for six of the 27 cancers that were studied. Most showed little or no difference between New Mexico and LAC. For some, such as oral and pharynx, the rates were uniformly lower in LAC. For much of the study, lung cancer in men was significantly lower in LAC. The spike in liver cancer, which is rather rare, occurred when there were four cases in 1981 and

1982. Previously there had been no cases since the beginning of the study in 1970.

12.3 THE EFFECTS OF COVARIATES OTHER THAN AGE

While the data were adjusted for age (see the Statistically Speaking box), they were not adjusted for any other variable. The presence of the Los Alamos National Laboratory means that the county has a number of high paying jobs. Wealth, as it turns out, has an effect on the risk of some diseases. For example, breast cancer is higher in more affluent areas.[4] This is likely due to wealthier women tending to have fewer children and having them at a later age. Since smoking rates are negatively correlated with education (i.e., smoking rates are lower for those with higher levels of education) lung cancer would be expected to be lower in areas with a highly educated populace, like LAC. Indeed, this is seen in Figure 12.2. Prostate cancer, which also tends to be uniformly higher in LAC than in New Mexico, is also higher in wealthier regions.[5] This is likely caused by more thorough cancer screenings in wealthier countries. LAC also experienced higher rates of melanoma of the skin. Although the research has yielded inconsistent conclusions, some studies point to higher melanoma risk in regions that receive higher levels of solar radiation. Since much of LAC is comprised of mesa tops that are approximately 7,000 feet above sea level, LAC would get more solar radiation, and if some of the studies are correct, this would translate into a higher risk for melanoma.

You can see that there are many factors that affect the risk of disease, so making comparisons between LAC and New Mexico as a whole is difficult. Is the difference in breast cancer due to the wealth of the county relative to New Mexico, or is there some environmental source causing this difference? Conversely, is the lung cancer rate in men due to lower smoking rates, or to an environmental factor? It is the epidemiologist's job to ferret out these effects and try to zero in on the real cause.

Most concerning, however, are the increases in brain cancer and thyroid cancer. Large spikes for both occurred in the late 1980s for no apparent reason. Brain cancer and thyroid cancer rates jumped by a factor of about four from the late 1970s to the late 1980s.

Another study[6] comparing cancer rates in LAC with the whole state of New Mexico reached similar conclusions. This report looked at cancer rates between 1970 and 1996 for 24 types of cancer. They looked at both cancer incidence (newly diagnosed cases of cancer that were not necessarily fatal), and mortality rates. For cancer incidence the lower confidence limit for the rate for LAC exceeded the rate for New Mexico as a whole. (This was the standard for statistically significant.) By contrast, for cancer mortality, the rate for New Mexico exceeded the upper confidence limit for LAC. Thus, cancer incidence was higher for LAC than for New Mexico, but the reverse was true for cancer mortality.

Among the 24 cancers studied, seven had statistically significantly higher incidence rates in LAC, and seven had statistically significantly lower incidence rates in LAC. For cancer mortality, only one type of cancer (breast cancer) had a statistically significantly higher rate in LAC than in New Mexico. Several had lower rates in LAC, but these were often for rare cancers that had zero cases.

The original concern was about brain cancer in LAC. Was this a signal of an elevated risk of brain cancer? Or was it just noise in a random system? By focusing on one type of cancer in one particular region we are looking at just one piece of a large puzzle. Such an analysis should consider the fact that multiple tests are being performed. Given that cancer occurs seemingly at random makes it difficult to detect a change in the cancer risk.

Notes

[1]See `www.cancer.gov/about-cancer/understanding/what-is-cancer`.

[2]See `www.lanl.gov/about/history-innovation/index.php`.

[3]See Athas, W.F., and C.R. Key (1993). Los Alamos Cancer Rate Study: Phase I. Final Report. Santa Fe, New Mexico: The State of New Mexico.

[4]See `http://abcnews.go.com/Health/CancerPreventionAndTreatment/story?id=1362570`.

[5]See, for example, `www.forbes.com/sites/peterubel/2017/12/22/why-living-in-a-rich-country-can-give-you-cancer/#77b82ca77288`.

[6]See `www.nuclearactive.org/docs/RTKAir.pdf`.

CHAPTER 13

Discovering the Cause of Yellow Fever

Y<small>ELLOW</small> fever is a viral disease that has plagued port cities for centuries. Symptoms for a mild case of yellow fever include fever, headache, muscle pain, nausea, and vomiting. In more severe cases, which occur in about one person in five, the symptoms include internal bleeding and liver and renal failure. Liver failure causes jaundice, giving rise to the name *yellow fever*. Internal bleeding in the gastrointestinal tract can cause the victim to vomit, ejecting what is described as coffee grounds. This is the source of the Spanish name for yellow fever, *vomito negro*. The fatality rate for the severe form of the disease ranges from 20% to 50%, though those who have survived yellow fever have lifelong immunity to the disease.

Yellow fever originally evolved in Africa about 3,000 years ago. It was carried to the New World in the 1600s, possibly through the African slave trade. Historically, the disease would appear in the spring or early summer, often after a ship had arrived in port. Invariably, there would be a delay of ten to fifteen days between the time an infected person arrived (usually by ship) and the time of the first local infection. The spread of the disease seemed haphazard, where the caregivers could remain disease free but those across the street having no direct contact with a sick person could become ill. It seemed like anyone in the vicinity of a sick person had the same chance of contracting yellow fever, regardless of contact.

A yellow fever outbreak in Philadelphia in 1793 killed 5,000 people out of a population of about 50,000. Today this 10% fatality rate is unfathomable; this would mean 150,000 deaths in the modern day Philadelphia which has a population of about 1.5 million. New Orleans, which saw alternating periods of high and low rates of yellow fever throughout the 1800s, experienced an outbreak in 1853 that killed nearly 8,000 people. Then, as the railroad system developed through the United States, yellow fever began to "ride the rails." For example, an outbreak in 1878 affected the Mississippi River valley as far north as St. Louis, with Memphis being the hardest hit city along the river, killing about 20,000 people. The first effort to build the Panama canal in the late 1800s, attempted by the French, was stymied by the high rates of disease, including malaria and yellow fever. The United States eventually took up the task when the French failed and the ability to contain yellow fever was a main factor in why the Americans succeeded.

This chapter is not about a particular outbreak investigation. Rather it is about the discovery of the method of transmission of yellow fever, where we will explore the key role that statistics plays in scientific investigations, particularly the clever and careful planning of experiments which were necessary to establish the cause-and-effect relationship. The statistical idea of having a control group for comparison plays a key role in the evolving sets of experiments conducted to determine how yellow fever is spread.

As with the transmission of cholera discussed in Chapter 1, which is spread by fleas, making the case that yellow fever is transmitted by mosquitoes took decades for a theory to accumulate evidence. Evidence, however, is not proof, and it took a dedicated team of Americans to doggedly pursue experiments that ultimately *proved* that mosquitoes carried the disease.

13.1 DR. FINLAY AND THE MOSQUITO THEORY

The mosquito theory of disease transmission dates back to the early 1840s to Louis-Daniel Beauperthuy, a French physician who practiced medicine in the Caribbean and in South America. He made a case for the *Culex faciata* mosquito (now called *Aedes aegypti*) being the source of transmission of yellow fever. Part of his case was to match the geographical distribution of the *Aedes aegypti* mosquito with the regions that suffered repeated outbreaks of yellow fever. Even though the regions overlapped quite a bit, this did not prove the connection. It could, of course, be the case that there is another cause that is present in the same types of climate in which the *Aedes aegypti* mosquito thrives. This was evidence, but not proof.

Statistically Speaking

Statisticians would say, "correlation does not imply causation." Correlation is a numerical measure that quantifies how closely two variables match. A positive correlation exists when two variables tend to either increase or decrease in tandem. A perfect positive correlation is represented by $+1$, while zero indicates no correlation, and -1 indicates a perfect negative correlation.

Two variables can be correlated without a causal relationship between the two. For example, if we look at national ice cream sales over the past ten years we will find that it is correlated to the number of drowning deaths over the same period. Of course, neither do drownings cause increases in ice cream sales nor does ice cream cause death by drowning. What is happening is that both ice cream sales and drownings increase in the summer and decrease in the winter, so there is a correlation between the two, but there is no causal relationship.

In the case of Beauperthuy's, comparison of the geographical distribution of the *Aedes aegypti* mosquito with the regions that suffered repeated outbreaks of yellow fever is an example of a positive correlation. While this positive correlation suggests

> some type of relationship between the *Aedes aegypti* mosquito
> and outbreaks of yellow fever outbreaks, it does not *prove* that
> mosquitoes cause yellow fever outbreaks.

Dr. Carlos Finlay (see Figure 13.1) was born in 1833 in Cuba and studied medicine at Jefferson Medical College in Philadelphia. Finlay began studying yellow fever and its causes in the late 1870s and he believed that the disease entered the body through the blood stream, which meant that some external agent was necessary to transmit yellow fever between people.[1] He noted:

> *It occurred to me that to inoculate yellow fever it would be necessary to pick out the inoculable material from within the blood vessels of a yellow fever patient and to carry it likewise into the interior of a blood vessel of a person who was to be inoculated. All of which conditions the mosquito satisfied most admirably through its bite.*[2]

In 1879, the United States sent a commission to Cuba to study yellow fever and Dr. Finlay assisted them. In the

Figure 13.1 Photograph of Dr. Carlos Finlay.[3]

early 1880s, Finlay began recruiting human volunteers. When he had twenty volunteers, he assigned five as experimental subjects, and the other fifteen as control subjects. The five were bitten by mosquitoes that had previously bitten sufferers of yellow fever. Four of the five developed yellow fever while none of the control subjects developed yellow fever.

Finlay presented his findings to the Havana Academy of Sciences on August 14, 1881. He presented a detailed description of the *Aedes aegypti* mosquito and how its geographical and elevational distribution correlated with the presence of yellow fever. He also presented the results of his experiments on human subjects. Realizing that his experiments were lacking an element of control, he admitted that "These tests certainly favor my theory, but I do not wish to exaggerate in considering as fully verified something which still needs further proof."[4]

At the end of Finlay's talk, there was silence: no questions, no comments. His revolutionary ideas were rejected, and he was the subject of ridicule. The flaw in Finlay's experimental design was that none of the subjects were in truly controlled circumstances. All were in Havana where yellow fever was endemic and it is conceivable that those who contracted yellow fever did so from a source other than a mosquito. This is an example of what statisticians would call a "confounding variable," something we will discuss in more detail in the next section. It would take another team and more experiments to provide more rigorous evidence.

13.2 THE YELLOW FEVER BOARD

The Spanish-American War was mostly fought in Cuba and lasted only a few months during 1898. While several thousand American troops died during the war, as is true in most wars only a small fraction died from combat; most died from diseases such as yellow fever, typhoid, or malaria. In 1900 American troops remained on the island of Cuba, and yellow fever continued taking its toll. Army Surgeon General George

Major Walter Reed   Dr. James Carroll

Dr. Aristides Agramonte   Dr. Jesse Lazear

Figure 13.2 Members of the Yellow Fever Board.[3]

Sternberg appointed a special commission of physicians to study the cause of yellow fever and whether it could be prevented. He appointed Major Walter Reed to lead the Yellow Fever Board, which included Dr. Jesse Lazear, Dr. Aristides Agramonte, and Dr. James Carroll. (See Figure 13.2.)

When the Yellow Fever Board began working in Cuba, there were two competing theories about the method of transmission of yellow fever. Finlay had espoused the mosquito theory, while others believed in the *fomite* theory. Fomites were believed to be organisms that lived in the soiled clothing and bedding material of yellow fever victims. After Finlay's experiments, Walter Reed and the Yellow Fever Board realized that only *controlled experiments* on humans would provide solid scientific evidence about the cause of yellow fever.

Reed said, "Personally, I feel that only can experimentation on human beings serve to clear the field for further effective work."[5] James Carroll wrote in the summer of 1900 that "The serious nature of the work decided upon and the risk entailed upon it were fully considered, and we all agreed as the best justification we could offer for experimentation upon others, to submit to the same risk of inoculation ourselves."[6]

Statistically Speaking

A *controlled experiment* changes only a fixed number of variables (often just a single variable) at a time, with all others kept as similar as possible, in order to isolate the results of those two variables. The idea is that if a change in one variable results in a change in the other, all else held constant, then there is strong evidence of a causal relationship between the two. In the 1920s, statisticians invented the field of *experimental design* to help researchers create experiments to most effectively learn about such phenomena. Sophisticated experimental designs allow for changing multiple variables simultaneously and controlling for extraneous effects.

In early experiments, Lazear took care of the mosquitoes, keeping each one in its own test tube. He allowed mosquitoes to feed on yellow fever patients at Las Animas Hospital in Havana and this was called "loading" the mosquito so that it might pass the disease to others. He then subjected himself to the first bite by a loaded mosquito on August 16, where his mosquito had been loaded by biting a yellow fever patient ten days earlier. Although unknown at the time, there is a period of about 4 to 14 days after a mosquito bites an infected person until the mosquito is able to transmit the disease. Because of this *extrinsic incubation period*, Lazear did not get sick.

Carroll was next. He said, "I acknowledge frankly that I was, in a measure, an unbeliever at the time I submitted to inoculation, for the evidence Finlay presented was far from convincing, chiefly because he worked in a hot bed of disease."[7] By "hot bed of disease" Carroll meant that Finlay's experiments were done in Havana, where the disease was common and so even if yellow fever follows a bite from a loaded mosquito, it is not possible to tell whether the bite caused the disease or whether the disease came from some other method of transmission.

Statistically Speaking

When it is impossible to determine which of two (or more) variables caused an outcome, the variables are said to be *confounded*. Here the effect of the mosquito bite and being in a high risk environment were confounded. Since Carroll had not been isolated from the presence of yellow fever, it was not possible to infer that the mosquito bite caused his disease.

On August 27, 1900, a loaded mosquito was allowed to bite Carroll's arm. Within two days, Carroll was feeling ill, and a few days later his temperature was 103.6 and he was jaundiced. He eventually recovered from the disease, but not until suffering the severe form of yellow fever and coming perilously close to death.

In the meantime, Lazear had found a volunteer: Private William Dean. He had been hospitalized at Columbia Barracks, just west of Havana, on an unrelated condition. Yellow fever was unknown in the vicinity of Columbia Barracks, which made Dean an ideal candidate because it would eliminate a confounding variable. Dean was bitten on August 31, 1900 by the same mosquito that bit Carroll and by September 5th he became ill from yellow fever.

Dean's case of getting yellow fever from the bite of a mosquito was the most compelling piece of evidence so far, including all of Carlos Finlay's experiments. He had come directly from the United States, so was unlikely to have had yellow fever earlier in his life. He spent his entire time (so far) in the tropics at Columbia Barracks, where there was no yellow fever. He then became sick after a mosquito bite from a loaded mosquito. This was the first undisputed case of yellow fever being experimentally transmitted by a mosquito bite. Dean was much younger than Carroll and in much better physical condition, and he recovered quickly.

The next to contract yellow fever was Jesse Lazear. He claimed to have been bitten by a mosquito on September 13, 1900 while loading mosquitoes at Las Animas Hospital in Havana. He said he mistook the species of the mosquito that landed on his arm to be something other than the *Aedes aegypti* mosquito. His notebook entry dated September 13, 1900 reads,

> Sep. 13. This guinea pig bitten today by a mosquito which developed from egg laid by a mosquito which bit Tanner – 8/6. This mosquito bit Suarez 8/30.[8]

Many now believe that Lazear had intentionally infected himself by being bitten by a loaded mosquito, and described himself as "this guinea pig" in his notebook. Lazear developed a severe case and died on September 25, 1900 at the age of thirty-four.

Major Walter Reed had been in the United States when these first experiments on humans were conducted. When he heard the details of the experiments, Reed became angry that they were performed under such uncontrolled conditions. After hearing of Lazear's illness, but before hearing of his death, Reed wrote to Carroll:

> *If you, my dear Doctor, had, prior to your bite remained at Camp Columbia for ten days, then we would have a clear case, but you didn't! You went just where you might have contracted the disease from another case. ... Unfortunately Lazear was bitten at Las Animas Hospital! That knocks his case out; I mean as a thoroughly scientific experiment.*[9]

Without using the term, Reed knew that the issue of confounding was present and that additional experiments under controlled conditions were needed. In October 1900, Reed approached General Leonard Wood requesting funding for further experiments, this time under more controlled circumstances.

13.3   BETTER EXPERIMENTS: CAMP LAZEAR

In order to better control for confounding factors, the Board decided to plan and conduct improved experiments. First, they moved to a new location which was free of yellow fever and that was dry with few places for water to accumulate (since mosquitoes lay eggs in standing water). The location was also relatively isolated, preventing too many people from coming and going, which would help eliminate the possibility that yellow fever was somehow introduced to the area by visitors. Reed named the place Camp Lazear.

At Camp Lazear, two small (20 × 14 foot) wooden buildings were constructed for two different experiments conducted from November 1900 through January 1901. In the first experiment, volunteers would be assigned to one of two *treatment groups*. In one group, volunteers would receive bites from loaded mosquitoes to test the mosquito theory of yellow fever transmission. In the other group, volunteers would be exposed to clothing and bedding from people who had yellow fever to test the fomite theory of transmission.

In the second experiment, volunteers would be housed in a single building and receive one of two treatments: exposure only to one side of the building that had mosquitoes or only to the other side of the building that was free of mosquitoes.

In modern terms, each experiment was a clinical trial with two "arms," where an arm refers to a group of subjects given a particular treatment. In the first experiment, the treatments were "exposure to loaded mosquitoes" or "exposure to contaminated bedding" and the design is shown in the top part of Figure 13.3. In the second experiment, the treatments were "exposure to loaded mosquitoes" or "no exposure to mosquitoes," the latter being a control group which would be exposed to

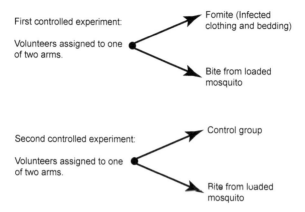

Figure 13.3 Arms of two controlled experiments to test the transmission of yellow fever.

the same environment in all other respects. (There was also a theory that yellow fever was transmitted via contaminated air). The design is shown in the bottom part of Figure 13.3.

The Yellow Fever Board used much of the money pledged by General Wood to recruit and pay volunteers. Volunteers received $100 for participating and an additional $100 if they contracted yellow fever. The Board actively sought volunteers from among Spanish immigrants and they were careful to exclude volunteers from regions that had seen yellow fever because having yellow fever confers lifelong immunity. In doing so, they were removing another potential confounding variable.

Experiment #1

In this experiment, volunteers in one arm were assigned to one of the buildings that was officially called Building 1 but unofficially referred to as the "infected clothing and bedding building." The investigators placed in this building the clothing and blankets that were soiled by the urine, feces, blood, and vomit from those known to have yellow fever. The building was then sealed so that mosquitoes could not enter and these materials were unpacked. The tight seal also meant that the stench from the clothing and bedding could not escape. Three volunteers stayed in this environment for twenty nights and none of them developed yellow fever.

In the other experimental arm, four volunteers received bites from loaded mosquitoes. Of these, none developed yellow fever, perhaps because of the extrinsic incubation period. However, later one of the volunteers, John Kissinger, was bitten again, and this time he developed yellow fever. Later, three additional volunteers developed yellow fever after bites by loaded mosquitoes.

Experiment #2

In the second experiment, as shown in Figure 13.4, Building 2 was fitted with a mosquito net to create a partition, preventing mosquitoes from passing back and forth between the two sides. The building was also sealed as much as possible to prevent mosquitoes from entering.

Initially one volunteer, John Moran, entered the mosquito side and was bitten several times by loaded mosquitoes that had been released on the mosquito side. Moran developed symptoms of yellow fever on Christmas day 1900. Later one

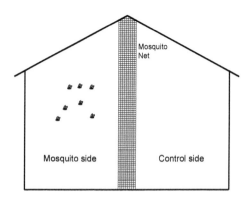

Figure 13.4 Camp Lazear Building 2 of the controlled experiments.

more volunteer was bitten by two mosquitoes, one on each of two consecutive days, but he did not develop yellow fever. In addition to the two men who were bitten on the mosquito side of the net, another two men slept in the control side for eighteen nights. They were in the same building, and breathed the same air as the two who were bitten, but they did not contract yellow fever.

Thus, by the middle of January 1901, evidence was mounting that bites by the *Aedes aegypti* mosquito was the method of transmission for yellow fever. The fomite theory was put to rest because none of those who slept in the feces and vomit soiled bedding developed the disease. Reed wrote to his wife saying that previous investigators:

> ... *were simply accepting the statements of others who had nothing on which to base such statements. A little careful testing of this* [fomite] *theory has served to knock it to smithereens. I thank God that I did not accept anybody's opinion on this subject, but determined to put it to a thorough test with human beings in order to see what would happen.*[10]

Reed had always credited Carlos Finlay with the pioneering idea behind the mosquito theory. In a letter to his wife on December 9, 1900, he wrote,

> *It was Finlay's theory, and he deserves great credit for having suggested it, but as he did nothing to prove it, it was rejected by all, including General Sternberg* [Army Surgeon General].... *I suppose old Dr. Finlay will be delighted beyond bounds, as he will see his theory at least partially vindicated.*[11]

This story of how the source of yellow fever transmission was discovered is really a story about how evidence regarding public health is gathered and evaluated. The experiments run by the Yellow Fever Board were really clinical trials, where one treatment was imposed on some subjects, and another treatment on other subjects, with confounding variables controlled as much as possible. As a result of these experiments, more than 30 men participated in the experiments, 22 developed yellow fever, but none of the volunteers died.[12] However, in the course of or related to the Board's experiments at least four died: Lazear, Clara Maass, a 25-year old nurse, and two Spanish immigrants, Antonio Carro and Cumpersino Campa.[13] One participant, John Kissinger, was incapacitated for life.

Statistically Speaking

One of the main aspects of clinical trials today is *informed consent*; that is, subjects who volunteer for a clinical trial must be informed of the risks and consent to the treatment. Most clinical trials today also involve *randomization* and *blinding*. Here blinding means that the subject (and usually the treating physician in a double-blind study) does not know which treatment the patient receives. While blinding was not used in the Yellow Fever Board's experiments, written consent was. At the time, consent was a revolutionary idea. Susan Lederer wrote that the consent used by the Yellow Fever Board "marked a significant departure in the history of human experimentation." [14]

Motivations for volunteering varied, however. To some the payment of $100 plus an additional $100 if he contracted yellow fever was sufficient inducement. ($100 in 1900 is worth approximately $2600 today.) For others, the motives were more altruistic, "solely in the interest of humanity and the cause of science." Some American soldiers refused to accept payment for their participation. Others were totally ignorant of the risk of getting yellow fever. For yet others, there was a certain amount of fatalism; they believed that moving to a place like Cuba would naturally put them at high risk for yellow fever. Catching it once would render them immune for life, and if they were going to get it, they might as well have it under the best medical care available.

Today, the spread of yellow fever is controlled by three mechanisms. One is the elimination, to the extent possible, of the *Aedes aegypti* mosquito. This is done by spraying and by avoidance of standing water that allows mosquitoes to lay eggs. This has other important health benefits, because the *Aedes aegypti* mosquito also carries other diseases, including Zika, chikungunya, and dengue. The second mechanism is vaccination. Max Theiler developed a vaccine called 17D in the late 1930s, for which he was awarded the 1951 Nobel Prize in Physiology or Medicine. The final mechanism is avoiding insect bites. This can be done by using mosquito netting, wearing clothing to protect skin, using insect repellent, and avoiding outdoor activity at peak hours (i.e., at dawn and dusk).

Based on the work of Dr. Finlay and Major Reed, the last yellow fever outbreak in the United States occurred in 1905, the last death attributable to yellow fever in Panama was in 1906, and yellow fever was eradicated from Cuba by 1908. Nonetheless, even today the World Health Organization estimates that forty-seven countries have regions that are endemic for yellow fever. Most (34) are in Africa, and the remainder (13) are in South America. WHO estimates that in 2013, there were between 84,000 and 170,000 severe cases, resulting in 29,000 to 60,000 deaths.

Notes

[1] Epidemiologists use the term *vector* to indicate an agent, often an insect, that transmits a pathogen from one person or animal to another.

[2] Chaves-Carballo, E. (2005). "Carlos Finlay and Yellow Fever: Triumph Over Adversity," *Military Medicine*, Volume 170.

[3] Source: U.S. National Library of Medicine, Digital Collections.

[4] Pierce, J.R., and J.V. Writer (2005). *Yellow Jack: How Yellow Fever Ravaged America and Walter Reed Discovered Its Deadly Secrets*, Wiley, p. 80.

[5] *Ibid*, p. 145.

[6] *Ibid*, p. 146.

[7] Carroll, J. (1903). *The Journal of the American Medical Association*, **41**, p. 44.

[8] Jurmain, S. (2013). *Secret of the Yellow Death: A True Story of Medical Sleuthing*, HMH Books for Young Readers, p. 48.

[9] Pierce & Writer, 2005, p. 156.

[10] *Ibid*, p. 185.

[11] *Ibid*, p. 182.

[12] Pierce, J.R. (2003). In the Interest of Humanity and the Cause of Science: The Yellow Fever Volunteers, *Military Medicine*, **168**, 857-863.

[13] Chaves-Carballo, E. (2013). Clara Maass, Yellow Fever and Human Experimentation, *Military Medicine*, **178**, 557-562.

[14] Lederer, S. (1997). *Subjected to Science: Human Experimentation in America Before the Second World War*, Johns Hopkins University Press, p. 21.

CHAPTER 14

Microcephaly and Zika

THE city of Recife lies on Brazil's Atlantic coast well north-east of the large cities of Rio de Janeiro and Sao Paulo (see Figure 14.1). This was the center of a strange phenomenon that began in the late summer of 2015. Babies were being born with abnormally small heads and brains. Their faces were usually the normal size and shape, but the small underdeveloped brains led to severe conditions, including blindness, deafness, seizures, hyperactivity, and others. Many of these babies died. It wasn't just a few cases; the rate was many times what it had been in previous years. Physicians were sometimes seeing several cases per day.

The condition is called microcephaly, and it occurs when newborn babies have a head that is significantly smaller than average. Operationally, it is defined as having a head circumference that is more than two standard deviations below the mean for the child's sex and age. The smaller head size is usually associated with underdevelopment of the brain and potentially devastating neurological conditions. See Figure 14.2.

Microcephaly is not unique to any part of the world and it is usually rare. The 2015 explosion in the number of cases in Brazil, particularly in Recife, led researchers to look for

Figure 14.1 Map of Brazil showing Recife where a large number of cases were reported in 2015.

a cause. Figure 14.3 shows the number of microcephaly cases across time in Brazil. The level was steady, with some inherent noise, at around 150 cases per year when the number jumped to 3,174 in 2015 and then to 8,092 in 2016. This jump is well beyond what would be expected by the background noise. In the sparsely populated and rather poor northeastern part of Brazil, the number increased from an average of about 40 cases per year to 876 cases. This is more than a 20-fold increase in the usual number and represented 90% of Brazil's total case count.

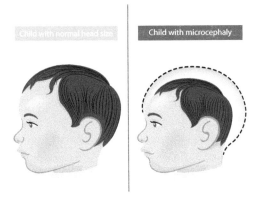

Figure 14.2 Comparison of a normal and microcephalic heads and brains. © Shutterstock Images.

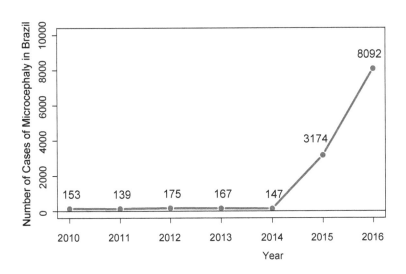

Figure 14.3 Microcephaly cases in Brazil from 2010 through 2016.

One of the first explanations was the Zika virus which had been virtually unknown in South America until 2015. It had been discovered in 1947 in the Zika forest in Uganda. The virus is carried by mosquitos, mostly but not exclusively by the *Aedes aegypti* mosquito. Mosquitos also carry other diseases, including malaria, yellow fever, dengue, chikungunya, and West Nile. Zika is also the only known mosquito-borne virus that can also be spread through sexual contact.

Zika is often a mild disease and it is estimated that 80% of those who have it will exhibit no symptoms. For those who do, the symptoms can include rash, fever, muscle/joint pain, and headache. For pregnant women, though, it seemed like the effect could be great. Since microcephaly cases surged after the Zika virus hit, public health officials suspected a cause-and-effect relationship between Zika and microcephaly. Once Zika became the suspected cause of microcephaly, other competing explanations began to emerge.

14.1 OTHER EXPLANATIONS

In his book *Zika: The Emerging Epidemic,*[1] Donald McNeil discusses many of the alternative theories that arose. One theory involved vaccines for rubella, also known as German measles, which is known to cause birth defects. But Brazil had not experienced an outbreak of rubella and there was no spike in rubella vaccinations. Also, rubella vaccines are used throughout the world, including the United States where it is part of the MMR vaccine, and there have been no great increases in microcephaly.

Another theory involved genetically modified mosquitos. It is true that a British company had bred male mosquitos that would pass on a trait that led to most males dying before becoming an adult capable of mating. Testing in Brazil was done in Piracicaba, 1,700 miles from Recife which is well beyond the range that mosquitos fly in a lifetime (about one mile).

A third culprit was pesticides and larvicides (which are used to kill mosquito larvae). Just as with vaccines, these pesticides are used worldwide and the steep rise in microcephaly occurred almost exclusively in Recife, Brazil.

Lastly, some had claimed that there was no outbreak of microcephaly at all. Clusters of diseases are to be expected even in the face of complete randomness. We saw this in the case of cancer clusters in Los Alamos in Chapter 12. Figure 14.3, though, makes it clear that microcephaly cases had surged in 2015 and 2016.

In early 2016, a study of 404 confirmed cases of confirmed microcephaly (out of a total of 4,783 reported cases) yielded only 17 babies who had tested positive for Zika. This suggested that Zika might not be the culprit after all. While this low percentage of microcephalic babies with Zika (17 out of 404 is just over 4%) it is also true that antibodies will normally destroy a virus in about two weeks. Many of the mothers had contracted Zika early in their pregnancy, giving their system plenty of time to get rid of the virus.

14.2 EVIDENCE

By early 2016 there were a number of competing theories about the cause of microcephaly. The first study that strongly linked microcephaly and Zika was a small-scale prospective cohort study. (This design had characteristics of the case-control study also.) A number of pregnant women in Rio were recruited into the study. Subjects must have been pregnant and have had a rash, which is one of the symptoms of Zika. Each was then given a Zika test and the cases were those women who tested positive and the controls were those who tested negative. Fifty-eight women remained in the study until they gave birth. The response variable was whether the baby had a "grave outcome," which included death in the womb, microcephaly at birth, and other issues. The results, summarized in Table 14.1, were striking. Among the sixteen women who tested negative there were no grave outcomes. Among the

Table 14.1 Preliminary results of prospective cohort study of Zika and microcephaly and other brain conditions.

		Grave Outcome		
		Yes	No	
Zika	Yes	12	30	42
	No	0	16	16
		12	46	58

42 women who tested positive for Zika, twelve babies had grave outcomes. This discrepancy of 29% versus 0% is highly statistically significant. In other words, if the probability of having a grave outcome for women who had had Zika was the same as the probability for those who had not, then we observed a very unlikely event. Because the chance of this is so small (much less than 0.0001) we are reluctant to believe the hypothesis of no difference.

This was a small-sample preliminary study, but the results seemed to be clear: women who had contracted Zika during pregnancy were more likely to give birth to babies with serious birth defects, including microcephaly. But women were not assigned at random; so there is still a possibility that there is one or more confounding variable that affects both the chance of contracting Zika and having a microcephalic baby.

A more complete study was published in *The New England Journal of Medicine*. This time, a total of 345 pregnant women who initially reported a rash were enrolled. Of these, 182 tested positive for Zika and 163 tested negative. After some dropouts, there were 125 in the Zika positive group and 61 in the Zika negative groups. The data are summarized in Table 14.2. The outcome was broadened to "adverse outcome" so there were a few more cases. Within those in the Zika cohort, 46% of pregnancies led to an adverse outcome, compared to only 11% of those in the control group, a result that is again highly significant.

Table 14.2 Results of prospective cohort study of Zika and microcephaly and other brain conditions.

		Adverse Outcome		
		Yes	No	
Zika	Yes	58	67	125
	No	7	54	61
		125	61	186

Statistically Speaking

An event A with probability p would in a long sequence of trials occur about $100 \times p\%$ of the time. We denote this probability by $\Pr(A)$. This is the relative frequency interpretation of probability. The *odds in favor* of the event A are defined as

$$\text{Odds in favor of event } A = \frac{p}{1-p}.$$

The *odds against* the event are $\dfrac{1-p}{p}$. For example, the probability of rolling a 🎲 on a six-sided die is $p = 1/6$, so the odds in favor of rolling 🎲 are

$$\text{Odds in favor of rolling } 🎲 = \frac{1/6}{1 - 1/6} = \frac{1}{5}$$

which is sometimes reported as 1 to 5 odds in favor. One interpretation of this is that there is one chance of the event happening against five chances of it not happening. The odds against are $\frac{5/6}{1/6} = 5$; in other words, the odds are 5 to 1 against.

Logistic regression is a method for quantifying the relationship between one or more predictor variables and a response variable that can take on only the value of zero or one. Such a variable is called **dichotomous** and is usually equal to one if some event A (such as having a baby with microcephaly) occurs and zero otherwise. The relationship between the predictor variables and the dichotomous response is that the log of the odds of A is linear in the predictor variables. Mathematically, this says that

$$\log \frac{p}{1-p} = \beta_0 + \beta_1 x_1 + \sum_{i=2}^{k} \beta_i x_i$$

where x_1 is equal to one if the subject was exposed to some factor (such as having Zika while pregnant) and $x_2, x_3, \ldots, x_k$ are possible confounding variables. It can be shown that the ratio of the odds for the one person in the exposure group ($x_1 = 1$) and a person in the nonexposed group ($x_1 = 0$) satisfies

$$\log \frac{\frac{p_1}{1-p_1}}{\frac{p_0}{1-p_0}} = \beta_1 + \sum_{i=2}^{k} \beta_i(x_{1,i} - x_{0,i})$$

where $x_{1,i}$ and $x_{0,i}$ are the predictor variables for the exposed and nonexposed persons, respectively. If both the exposed and nonexposed persons had the exact same values for all covariates $x_2, x_3, \ldots, x_k$ (i.e., the possible confounding variables), then the log odds would equal the constant β_1. This means that all things being equal, except for the exposure variable, the log odds of having the disease will be β_1. Note also, that if the log odds is zero, then the odds will be one, which tells us that the odds of getting the disease are the same for both the exposed and nonexposed groups. If the odds ratio, which is equal to $\exp(\beta_1)$, is greater than one, then the odds of getting the disease is higher in the exposed group than the nonexposure group.

In a case-control study[2] conducted at about the same time, researchers in Recife (the epicenter of the microcephaly crisis) selected cases of microcephaly and attempted to match each individual case with two controls. The controls were selected as follows. For a full- or post-term delivery, the investigators selected the next two eligible neonates who were born after 8:00 am the day after the birth of the case, and who were also full- or post-term. Similarly, for early preterm births of cases (defined as less than 34 weeks' gestation), the controls were selected to be the next two eligible births who were less than 34 weeks' gestation. An analogous procedure was done for cases born between 34 and 36 weeks' gestation. For various reasons, there were not always two controls for each case. Since this is a case-control study, we must compare the rates of Zika between the cases and controls. By contrast, in a prospective cohort study we would compare the rates of microcephaly between the Zika-positive and Zika-negative groups. The results were

Table 14.3 Results of a case-control study on the association between Zika and microcephaly.

		Microcephaly		
		Cases	Controls	
Zika	Yes	32	0	32
	No	59	173	232
		91	173	264

again astonishing. Among the 173 controls, there were no cases of Zika. Among the 91 cases of microcephaly there were 32 cases of Zika. The data are summarized in Table 14.3.

Researchers followed up on this raw calculation by applying logistic regression, allowing them to account for some other possible confounding variables. They accounted for smoking during pregnancy, skin color, and whether the mother received the Tdap (tetanus, diphtheria, and pertussis) vaccine while pregnant. The odds ratio adjusted for these variables was 73.1 with a confidence interval of $(13.0, \infty)$. This means that the odds that the mother of a microcephalic child had Zika were 73.1 times that for the mother of a nonmicrocephalic child. Among the 13,624 babies screened for this study, there were 101 cases of microcephaly, for a rate of 74 per 10,000 births. This is many times the rate of 6 per 10,000 experienced in the United States.

14.3 OTHER POSSIBLE CAUSES

The case that Zika causes microcephaly had been pretty well established by early 2017, but there were still some unanswered questions. Microcephaly rates were highest in northeastern Brazil, even though Zika had spread through much of the country. Also, Zika had been widespread in French Polynesia, a collection of islands in the south Pacific midway between Australia and South America, in 2013 and 2014. There was an outbreak of microcephaly in French Polynesia that was discovered only after the outbreak in Recife, Brazil. Eight microcephaly cases occurred in French Polynesia. An outbreak in Colombia led to a fourfold increase in the rate of microcephaly, compared to Brazil which experienced somewhere between a ninefold and a twentyfold increase. So why was the outbreak in Recife, Brazil, so bad?

Researchers at the London School of Hygiene and Tropical Medicine[3] have suggested that other diseases could be acting together with Zika to cause the disease. They point out that rates of chikungunya, a serious and painful but rarely fatal disease, were also high in northeastern Brazil at the same time Zika was spreading. Dengue fever is also a disease that is spread throughout most of the region. It might be that coinfections with two or more diseases may increase the risk of microcephaly. The researchers also found that poverty was an indirect risk factor for microcephaly. A possible causal diagram is shown in Figure 14.4.

In the previous chapter we saw that carefully designed experiments demonstrated that yellow fever was acquired by the bite of a mosquito that had bitten an infected person. In this

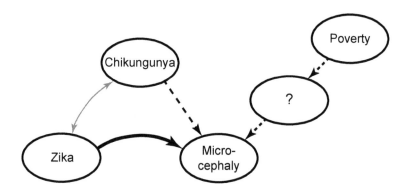

Figure 14.4 Possible causal diagram for microcephaly. Thick lines with a single arrow indicate causation. Thin lines with double arrows indicate an association that is not necessarily causal. Dashed lines indicate possible, but unproven, causal relationships.

chapter, we have looked at microcephaly, which seems to be caused by the mother's infection with Zika during pregnancy. Recent research suggests that other factors may work together with Zika to increase the risk of microcephaly. Further data from South America, the Caribbean, and possibly parts of North America will help to answer this question.

Notes

[1] McNeil, Donald G. (2016). *Zika: The Emerging Epidemic.* WW Norton & Company.

[2] de Araújo, Thalia Velho Barreto, et al. (2016). Association between microcephaly, Zika virus infection, and other risk factors in Brazil: final report of a case-control study. *The Lancet Infectious Diseases* **16(12)**, 1356-1363.

[3] Campos, Monica C., et al. (2018). Zika might not be acting alone: Using an ecological study approach to investigate potential co-acting risk factors for an unusual pattern of microcephaly in Brazil. *PloS one* **13(8)**, e0201452.

CHAPTER 15

In Conclusion

IN Chapters 8 through 10 we looked at investigating outbreaks of communicable diseases (Nipah, smallpox, and syphilis), and in Chapter 11 we looked at an intentional infection using anthrax. These examples applied many of the steps described in Chapter 7 that epidemiologists use to investigate outbreaks. In Chapter 12 we looked at an investigation regarding cancer, a noncommunicable disease. Finally, in Chapters 13 and 14 we studied the experimental designs that epidemiologists and statisticians use to discover the cause of a disease.

We looked at seven examples of disease outbreaks that used the epidemiological approaches described in Chapter 7. Four of the seven were contagious diseases (Nipah virus, smallpox, syphilis, and yellow fever). One example, microcephaly, is a consequence of a contagious disease, Zika. Cancer is likely caused by environmental factors. Finally, anthrax is caused by an agent purposely spread by human intervention.

A key focus of these statistical methods is on the idea of separating a disease signal from noise in data. This could be at an individual level in which a doctor seeks to reach a diagnosis based on laboratory tests that are imperfect for determining whether an outbreak is occurring amid the normal increases and decreases of disease incidence in a population. In many ways, it comes down to the question of whether some observed uptick in disease incidence today is the start of an increasing incidence trend or just a "blip" that will then drop back down tomorrow or the next day. The common theme among these is how scientists used data and statistical methods to guide them in their search for the cause of a disease outbreak.

A critically important aspect of disease surveillance is early detection: catching an outbreak as early as possible in order to mitigate the impact of a pandemic. To do this, particularly when looking at non-specific syndrome data, the challenge is developing and properly employing sophisticated statistical methods that are good at identifying increasing disease incidence trends while minimizing the chance that a blip is incorrectly identified as a trend. Here statisticians are the key players, continually developing new methods in order to improve performance.

15.1   FURTHER READING

For those who would like to learn more, we recommend the following for further reading.

1. Brookmeyer, R., and Stroup, D.F. (2004). *Monitoring the Health of Populations: Statistical Principles & Methods for Public Health Surveillance*, Oxford University Press, Oxford, UK.

2. Dobson, M. (2007). *Disease: The Extraordinary Stories Behind History's Deadliest Killers.* Quercus.

3. Dworkin, M.S. (2010). *Outbreak Investigations Around the World: Case Studies in Infectious Disease Field*, Jones and Bartlett, Sudbury, MA.

4. Fricker, Jr., R.D. (2013). *Introduction to Statistical Methods for Biosurveillance, with an Emphasis on Syndromic Surveillance*, Cambridge University Press, New York, NY.

5. Friis, R. H., and Sellers, Thomas A. (2014). *Epidemiology for Public Health Practice*, Jones and Bartlett, Sudbury, MA

6. Gigerenzer, G. (2002). *Calculated Risks: How to Know When Numbers Deceive You*, Simon and Schuster, New York.

7. Guillemin, J. (2005). *Biological Weapons: From the Invention of State-sponsored Programs to Contemporary Bioterrorism.* Columbia University Press.

8. Hayes, J.N. (2005). *Epidemics and Pandemics: Their Impacts on Human History*, ABC Clio, Santa Barbara, CA.

9. Kass-Hout, T., & Zhang, X. (Eds.) (2010). *Biosurveillance: Methods and Case Studies*. CRC Press.

10. Khan, A. (2016). *The Next Pandemic: On the Front Lines Against Humankind's Gravest Dangers*. Public Affairs.

11. Lawson, A. B. (2018). *Bayesian Disease Mapping: Hierarchical Modeling in Spatial Epidemiology, Third Edition*, CRC Press, Boca Raton, FL.

12. Lundberg, K. (2009). The Anthrax Crisis and the U.S. Postal Service, in *Managing Crises: Responses to Large Scale Emergencies.*, A. M. Howitt and H. B. Leonard, Ed., CQ Press, Washington, DC.

13. M'ikanatha, N. M., Lynfield, R., Van Beneden, C. A., & de Valk, H. (Eds.) (2008). *Infectious Disease Surveillance*. John Wiley & Sons.

14. McNeil, D. G. (2016). *Zika: The Emerging Epidemic*, Norton, New York.

15. Mehra, A. (2009). Politics of Participation: Walter Reed's Yellow-Fever Experiments. *AMA Journal of Ethics*, Volume 11, Number 4, pp. 326-330.

16. Pierce, J. R. and Writer, J. (2005). *Yellow Jack: How Yellow Fever Ravaged America and Walter Reed Discovered Its Deadly Secrets*, Wiley, Hoboken, NJ.

17. Quammen, D. (2012). *Spillover: Animal Infections and the Next Human Pandemic*. W.W. Norton and Company.

18. Yamada, I., & Rogerson, P. (2008). *Statistical Detection and Surveillance of Geographic Clusters*. Chapman and Hall/CRC.

19. Zelicoff, A. P. (2003). An Epidemiological Analysis of the 1971 Smallpox Outbreak in Aralsk, Kazakhstan. *Critical Reviews in Microbiology*, 29(2), 97–108.

20. Zelicoff, A. P., and Bellomo, M. (2005). *Microbe: Are We Ready for the Next Plague?* AMACOM Div American Management Association.

Index